KT-524-118

How to be Human: The Manual

A MONK, A NEUROSCIENTIST AND ME

PENGUIN LIFE

PENGUIN LIFE

UK | USA | Canada | Ireland | Australia
India | New Zealand | South Africa

Penguin Life is part of the Penguin Random House group of companies
whose addresses can be found at global.penguinrandomhouse.com.

First published by Penguin Life 2018
Published in this edition 2018
001

Copyright © Ruby Wax, 2018

The moral right of the author has been asserted

Set in 11.04/13.86 pt Dante MT Std
Typeset by Jouve (UK), Milton Keynes
Printed and bound in Great Britain by Clays Ltd, Elcograf S.p.A.

A CIP catalogue record for this book is available from the British Library

ISBN: 978–0–241–29475–8

www.greenpenguin.co.uk

Penguin Random House is committed to a sustainable future for our business, our readers and our planet. This book is made from Forest Stewardship Council® certified paper.

How to be Human: The Manual

C016690698

With special thanks to Ashish Ranpura for his outrageous knowledge of neuroscience and Buddhist monk Gelong Thubten for his wisdom of the mind and great sense of humour. Ash is funny too.

Also, with great gratitude to my editor (and wonderful husband) Ed Bye, and to my other editor (not my husband), Joanna Bowen.

And thanks to my kids Maddy, Max and Marina for not becoming crackheads. And making me so happy.

Contents

Thubten, me and Ash

Preface

After writing my last book I said to myself, never again. It's like having a baby: you're in such pain during the birth, all you want to do is chew your arms off; it's the same with writing a book, except you're dilated for more than a year. But when it finally comes out and it thrives (hits number one on that bestseller list), oh my God, all you want to do is get fertile and do it again. So here it is, my next baby.

IN THE BEGINNING WAS THE WORD . . . and then my book. I'm going to start where we all started, back in the swamp and with a fine-toothed comb find out exactly what happened to make us who we are today. (I spoke a little about this in *Frazzled* but am now digging deeper). Are we everything evolution dreamed we'd be? If not – who can we blame? Not that we can do anything about it, but if there's life on other planets, they may learn from us. The human race is a miracle when it comes to survival. If you're alive and reading this book right now, you're a bona fide, gold medal winner in the 'Evolutionary Hunger Games' where you had about a trillion to one chance of not being born a frog. This should make us the happiest species alive, but we're not. We spend our time on earth in a constant hunt for contentment, and as far as that goes we're in a holding pattern. So what's going to happen and where are we going?

You can't stop the future from arriving, no matter what drugs you're on. But even if nearly every part of us becomes robotic, we'll still have, fingers crossed, our minds which hopefully we'll be able to consciously upgrade, making us more human and less machine. You only need a mind to practise mindfulness and compassion skills (no fingers or toes necessary) and no amount of titanium in the world can give you those qualities. Mindfulness isn't for everyone but from my own experience and according to scientific evidence, mindfulness rules.

Maybe in the future someone will invent a wearable mental Fitbit which can help us achieve insights and awareness but it ain't here yet.

Since writing the last book, *A Mindfulness Guide for the Frazzled*, I've practised every day along with every excuse in the world not to do it. It's a constant struggle, but somehow, I squeezed it in, the result being I'm happier, calmer (except when I'm handed a traffic ticket; then I'll go for the jugular), and more able to focus my mind where I want it, when I want it and to me that's a big part of happiness.

Most importantly, it's helped me sense a depression coming before it hits. This doesn't mean I dodge it but now I'm ready for it. When I sense the tiny, far-off footsteps of despair, I batten down the hatches, swiftly unplugging from any contact with the rest of the world, both onscreen and in person, giving me a chance to cold-turkey off my addictions to emailing, needing to be liked by everyone, even people I don't like and worrying about what's going to nix us next – North Korea or too much salt?

I'd say I'm different since the last book, but my insatiable desire to know everything about everything remains, and I don't think that's such a bad thing. Luckily, I found a brilliant neuroscientist and a Buddhist monk to help answer some of my niggling questions. I figured the monk could explain how our minds work and the neuroscientist could tell me where it all goes on in the brain.

After living and breathing for what seems eternity with the monk and the neuroscientist, it feels like we're an old married threesome. We harangue, nag and bitch but our relationship continues to flourish because we can make each other hysterical. I might say to the monk, 'That's way too Buddhist-sounding. I don't want to hear another "B" word out of you tonight.' And he'll flip me back a 'That's two thousand years of wisdom, sweetheart, swallow it.' At one point, we were thinking of writing another book, combining our separate wisdoms, called 'Act Like a Buddha, Think Like a Jew'. When the neuroscientist starts cocking his high-IQ feathers at us, the monk and I tell him to start speaking in human tongue or we'll cut him out of the book.

I've been squeezing them dry for more than a year, leaving them empty and exhausted, but I got a book out of them, and that's all that counts. At the end of each chapter, I let them out of their cages to riff.

Chapter 1 Evolution

As I said earlier, it's those ancient whispers that have instilled in us a drive to be top gun or at least top gun's best friend. They made us survive as a species but, individually, they've made us miserable because we feel like we're in some race but we don't know what for? By showing you how we evolved and why, you'll come to realize that you are not your fault. Evolution did it. What a relief.

Chapter 2 Thoughts

Why do we have them? And why, oh, why are they so bitchy? My hope is that in this chapter you'll come to understand that your thoughts aren't who you really are. If they were, who is the one observing them? Once you understand this then you can pick and choose which thoughts to use and which to lose and that is the Yellow Brick Road to happiness.

Chapter 3 Emotions

Emotions are like the old joke they say about wives: you can't live with 'em, you can't live without 'em. The reason we have 'em is to help us navigate our lives; informing us what we like, don't like and why you chose to buy this book over all others (a huge thank you for that). Without emotions, we're zombies. So, really, it's not, 'I think therefore I am,' it's more like, 'I feel therefore I am.'

Chapter 4 The Body

Many of us don't think that our brains have any relationship to our bodies; they're strangers in the night. Some of us (like me) find the body to be an irritating piece of skin I have to drag around under my neck, like an old wedding train. The fact is, the brain and body are in constant communication, each influencing the other. If you think happy, your body is happy and vice versa.

Chapter 5 Compassion

So much social media and we're more isolated than ever before, partially because we're not talking to each other face-to-face, only face-to-screen.

In this world, few of us have the time for compassion, being so busy with such a tight schedule, but we're going to need it if we want to survive as a species, not forgetting that it's the glue that makes our lives worth living.

Chapter 6 Relationships

Who's got that one right? We're all caught with conflicting impulses, a crossfire of desires which tear us in opposite directions. The sex god or the safe one? This has always been the female dilemma: choosing between brute or good guy (see *Wuthering Heights* or *Gone with the Wind* or any episode of *Friends*). Many of us grew up with the delusion that someday, our prince will come. This explains why many of my friends are miserable today, because they never found one. We're so used to upgrading things like our iPhones as soon as they get old, we don't think twice about it, we dump them. Many people I know are now on iWife4 or iHusband8. The motto being, if it's new, it's better.

Chapter 7 Sex

You'll love this one.

Chapter 8 Kids

You are the sculptor of your child's mind. Every look, reaction and gobbledegook that spills out of your mouth, will influence who they'll ultimately become. Before you panic and reach for the Xanax, you can still learn to change your behaviour, thoughts and emotions to give your kids a better chance of becoming resilient, balanced and basically a better human being. You can teach an old dog new tricks.

Chapter 9 Addiction

Throughout history, there's always been something to be addicted to, but these were substances you could chew, smoke or snort; physical addictions. Now, we're addicted to eating, gambling, shopping, sex, the phone . . . the world has become an endless buffet of temptations. These days it's not just physical, we're also addicted to our obsessive thoughts. So the idea is, if we change our thoughts, we can change our addictions.

Chapter 10 The Future

I only hope that, whatever apps or robotic appendages we end up with, we'll still be able to look inside ourselves and have some awareness of our thoughts and feelings. The danger is that we might live life floating from one hit of pleasure to the next, from experiencing sex with blowfish to being able to read the minds of trees, which will only lead to a craving for more, and there won't ever be enough toys on earth to make you feel full.

Chapter 11 Mindfulness Exercises

For all the above topics, the monk and I will provide accompanying mindfulness exercises. All the practices are like using barbells in the gym, but with these weights you're making your mind more focused, more flexible, more aware, less distracted, less addicted, faster, healthier and above all, more compassionate.

Chapter 12 Forgiveness

We're all capable of forgiving but in our world, like compassion, with so many things to do and people to be better than, it's hard to slot it in. Only if we can forgive ourselves will we be able to forgive others. Rather than always finding someone to blame for our discontent, we might be able to finally negotiate and relate to people 'not like us' by seeing them as 'just like us'.

I'd now like to introduce the monk and the neuroscientist without whom I couldn't have written this book.

The Monk

Gelong Thubten became a monk at the age of twenty-one, at Kagyu Samye Ling Tibetan Buddhist Monastery in Scotland. I didn't know this until recently, but 'Gelong' is a title meaning life-long senior monk. I thought it was his first name. Thubten is now forty-six and teaches mindfulness at companies such as Google, LinkedIn, Siemens and many other organizations around the world. He also trains school kids and medical students in mindfulness and, more importantly, worked on this book with me.

A little about his childhood: his father, who is English, made a fortune as a successful computer programmer, and his mother is the well-known Indian actress Indira Joshi from *The Kumars at No. 42*. They are now divorced. At six, Thubten ran away from home, planning to hitch-hike around the world. He only made it to the end of the street, where they found him holding a globe of the world and a box of Kleenex. His mother took him home.

At school, the other kids thought Thubten (pronounced 'toop-ten') was highly intelligent and had no friends; they thought he was nerdy. What they didn't know was that, at night, he and his English teacher from school played music in venues across London. Thubten was a jazz pianist, tinkling the ivories while his teacher belted out classics like 'Summertime' and 'The Girl from Ipanema' in a smoky red dress. Thubten was fourteen at the time but pretended he was twenty-one; he slicked his hair back and wore a dinner jacket to look older. The pair planned to take their act on to cruise ships. No one at school knew about his double life.

Thubten went to Oxford University to study English literature but finally 'blew it out of the water' to pursue his dream and become an actor. He got an agent in London who sent him out to audition for parts. Most of them had the word 'Buddha' in the title: *The Buddha of Suburbia*, *Little Buddha* . . . He didn't get any of those roles, so decided to move to New York to pursue his dream. He lived an incredibly wild life as a budding actor but, eventually, after two years, he burnt out, coming close to having a heart attack. That was when he had the wake-up call that made him re-evaluate his life.

So, at twenty-one, Thubten went to a Buddhist monastery to get his head straightened out. Four days later, he became a monk. Everyone he knew, including his parents, were very surprised. First, he became a temporary monk for a year to clean up his act. His plan was to go back to New York after the year, to his old life. It never happened; he stayed for life.

After a year he went on a strict nine-month solitary retreat where he could only eat one meal and drink water on alternate days. He spent five months of it in total silence, practising meditation for twelve hours a day. Later, he went on an even more intensive retreat on a secluded island off the coast of Scotland, this time lasting four years, again with five months in total silence. At times, he says it became too much; he was holding on by his fingernails, sometimes literally clutching his seat, battling with his mind for two and a half years, facing his demons, and then, suddenly, there was a shift and he gave into the practices. He says he felt himself relax and make friends with his mind, and now knows how to hit a mental switch and access feelings of peace and happiness.

Thubten was trained by Akong Tulku Rinpoche (a seriously big player) who took him around the world, showing him how to speak for large audiences. As well as teaching Buddhist philosophy, Thubten has been a pioneer in teaching mindfulness in prisons, hospitals, schools, charities, universities, drug rehabilitation centres and companies, starting twenty years ago before it became popular. In those days it was always called meditation and only became known as mindfulness in the last ten years. When Thubten teaches, he keeps religion out of it, focusing on the breath, body and compassion. As a monk, Thubten doesn't get paid, but the big companies give donations which he uses to build retreats and mindfulness centres that help thousands of people.

Recently, he was asked by Disney to be the mindfulness advisor on the set of the movie *Dr Strange*, and he taught Tilda Swinton and Benedict Cumberbatch mindfulness between takes. (Both were already into it.)

I met Thubten at a conference in Sweden, loved him on sight and he now stays in my house when he's teaching in London. I call him 'my air freshener' or 'the human smudge stick'. We know how to make each other laugh until one or both of us fall on the floor.

The Neuroscientist

Ash Ranpura is a clinical neurologist and a neuroscientist. He received his Bachelor's degree in molecular neurobiology at Yale, his MD in clinical medicine at the Medical College of Ohio and worked on his PhD in cognitive neuroscience at University College London, where he also taught Master's courses in statistics and research methods. He conducted research on learning and autism at the University of California, San Francisco, before returning to Yale to complete his speciality training as a consultant in neurology. As a doctor, he sees patients with unusual disorders that fall between the disciplines of neurology and psychiatry, and he has treated everything from brain parasites to hysterical paralysis. Ash has published academic papers on topics ranging from growing slug neurons on glass and whether fish can count, to the prevention of communicable disease in Bangladesh and the use of medication in Alzheimer's. (I asked him, and he said fish could count but only up to about eight.)

I don't think I have to say that he's very smart, and I was intimidated when I met him. Now I've got to know him and I see his flaws loud and clear – too many to mention, so I'm not so scared of him any more. He was born in Dayton, Ohio, which he says was once home to the Wright brothers and is now the home of the twenty-four-hour pancake-breakfast special. His mother, who was born in Mysore, India, was recruited to the US for a job in internal medicine and anaesthesiology because of a dearth of qualified doctors in rural America. She met Ash's father (also a doctor of anaesthesia) at a hospital in Ohio (I'm guessing over someone they had just knocked out before surgery). Ash says his mother was overprotective and when he was a child she would do 'drive-bys' at his school and friends' homes to keep tabs on him. Later, she tried to stop him from dating, hoping for an arranged marriage. It didn't happen.

Ash inherited both his mother's and his father's intelligence; he never got a grade below an A at school. He says, 'Getting a B was an Asian F. Straight A's were the basic expectation.' Though his mother was highly educated, she believed in omens, saying things like, 'The cactus always blooms before your sister comes home,' or 'A bird chirping three times means your dead grandmother is watching us.' Ash describes her premonitions as 'Hallmark card meets witchcraft'. He assumed this mixture of the mystical and the medical was normal.

From the earliest age, Ash remembers that all his parents talked about were their plans for death. They made him promise at the age of three never to put them into a nursing home and not to take 'drastic measures' to save their lives. Drastic measures seemed to include taking anything for illness other than turmeric and warm milk. They had seen too many people end their lives suffering in hospital, and thought it would be wrong if they lived too long and wasted the family money. Ash said the first words he learned were 'Pull the plug.'

His father died suddenly of a heart attack, so the question of pulling the plug didn't come up. He was cremated, and his mother, always frugal, declined an urn and put the ashes in a cardboard box. One random day, she decided to burn the box and the ashes in a fire pit at a park in Cleveland; she was going to roast some corn for dinner anyway. Because things were running late (her sisters were coming for dinner), she thought she'd get everything going at the same time. She said a quick prayer and roasted the corn on top of the burning ashes of the father. They went home and Ash's sister made salsa out of the roasted corn and peppers. When his aunt found out about the barbecue/funeral, she left the house in a fury. But Ash's mother, maybe because she's a doctor, felt she was simply being practical.

She is eighty years old now, recently got a degree in psychiatry and still works at a hospital, sleeping there several nights a week.

Ash is married to Susan Elderkin, a successful novelist whom he met in London, and they have a son, Kirin, who is probably going to be a genius.

1
Evolution

What Exactly is This Thing Called Evolution?

Let me clear up something about evolution. I was taught in school that when we evolved as a species it meant we were improving the whole time; each generation developing more advanced features. I've found out this is a common misconception. Evolution doesn't mean the species gets better, it just means we become better adapted to the environment, sometimes at great cost. What works for us as far as surviving is concerned also might work against us. One example is us getting up on all twos, great for hiking but the downside is we get backache. Had we remained in crawl position, we'd be fine and chiropractors would be out of business.

People today may be living longer than they did before, but it's not clear that people are living better, or that in one hundred years from now they'll be living more improved lives than they are today.

What Went Wrong?

Evolution did a fine job helping us adapt to changes in the weather and the dodging of dinosaurs. Full points for helping us out with survival, but not so many for helping us figure out what we're supposed to be doing here on planet Earth. It's this sense of unrest, this nagging feeling that we're supposed to find some meaning, that makes us (especially existentialists) very, very unhappy. Baboons are still going around having the time of their lives while we're tearing out what little hair we have (compared to baboons) trying to suss out why we don't feel good enough. We're at the top of the food chain, for God's sake, what's to feel not good enough about? Even grasshoppers don't have low self-esteem (I'm guessing).

We can deal with danger but, in the face of envy or comparison, we're helpless. You can't club those emotions like you can a predator; they aren't a physical entity and you'd just be clubbing yourself to death. When we get trapped by envy, it grows into rage, which can grow into illness, addiction and, eventually, mental disorders, especially for kids.

It's a miracle evolution worked at all. What were the chances that, out of some stardust, we would make it through to now with our full set of teeth intact? As far as we know, no other planet managed to pull this off. They haven't made a single cell of anything interesting, while we've already sold 12 trillion McDonald's burgers.

Astronomer Fred Hoyle wrote, 'To imagine that a human being could emerge by random chance in the universe is like trying to imagine a hurricane blowing through a junkyard and creating a 747.'

We give ourselves such a hard time for things that are out of our control. For me, this news was a revelation; the fact that I am not my fault but merely a player in the DNA legacy has done wonders in helping me stop being so self-critical. My addictive drive to achieve, whether it's getting someone at a party to like me (who I'll probably never see again and/or don't even like) to saying, 'Of course I can write a book in three months' (and ending up institutionalized from the pressure). I now know that this drive isn't something I'm doing on purpose to torture myself; it's not just my condition, it's the human condition. It's something that's been passed down by our Palaeolithic forebears for survival's sake to keep us striving for rewards. Hurrah! I don't need to be absolved by any shrinks, priests or rabbis; human history is the culprit.

Our Relations

Why are we so hard on ourselves when, in our evolutionary timescale, we're still in our infancy? Here are a couple of facts to show that we as *Homo sapiens* are still a work in progress and not at all as cutting edge as we like to think. We share 98 per cent of our DNA with great apes, and about 90 per cent with mice. And it gets worse: we share 30 per cent of our DNA with yeast. I heard that there's a T-shirt with the slogan, 'You share 25 per cent of your DNA with bananas. Get over yourself.'

I read recently that researchers discovered a wrinkled sac-like body about one millimetre long that's all mouth and no anus (I'm not making this up), so food goes in and comes out of the same orifice. It is called Saccorhytus coronarius. It turns out it's related to us, even though it existed 540 million years ago and is now extinct. So there's another reason for us not to get up on our high horse.

Darwin wrote, 'Why is thought, being a secretion of the brain, more wonderful than gravity, a property of matter? It is only our arrogance and our admiration of ourselves.'

The History of Us in a Nutshell

We began as a single-celled piece of protoplasm sticking to a rock (it was a pathetic sight). We remained in that state for millions of years, then we advanced to algae (not an impressive leap). Later, we went through our fungal phase (yes, you are related to the mould on last week's sandwich and old yogurt). Next, we became parasites, and moved on to being jellyfish, to worms, to jawless fish, to sharks (Wall Street stage). Our 'fish out of water' amphibian period ended when we replaced our fins with feet and crawled out of the water. After that, there was no stopping us. At this point in our early mammal days, we looked a little like hamsters and spent our time dodging dinosaur feet. Then, accidentally, a meteorite fell down. Those of us who survived continued to develop. The rest were squished.

Somewhere between 40 to 25 million years ago we went from being apes to being orang-utans to being chimps. About 6 million years ago we went bipedal and homo. (Not that kind of homo – stay focused.)

Only for the last 200,000 years have we been modern humans: *Homo sapiens*. 'Sapiens' translates roughly into 'thinking about thinking'. We were, and are still, the newest kids on the block. This doesn't mean we're now flawless. We didn't dump all our old primitive equipment; those reptilian parts are still alive and well within our brain. So that's where we are today: part savage, part genius. That ancient region left over from about 300 million years ago had its uses, endowing us with the ability to breathe, swallow, hump, sneeze (the basics). *Eat, Pray, Love* would never have made it as a bestseller back then. *Eat, Fuck, Kill* might have been the bigger hit.

Our mammalian brain is about 100 million years old (giving us some emotional range and the ability to bond). About 200,000 to 500,000 years ago, an area of our brain known as the neocortex had a growth spurt, giving us the ability to plan, to self-regulate, to control our impulses and become aware of ourselves. With this more advanced part of our brain, we learned to speak, to use symbols, solve problems and imagine the future. The downside was we started to worry and ruminate about 'what if?' scenarios, not to mention the mother of all worries – knowing that we're all going to die – which all adds up to make us a very jittery race.

Trade-offs

It seems that, in evolutionary terms, every time we take one step forward, we take a multitude back. These evolutionary trade-offs don't just happen to humans but to all living things. An animal which I think got one of the worst deals is the giraffe. It never evolved claws, sharp teeth or a hard shell and needed some feature to avoid getting extinct-ed (I know it's not a word, but I'm using it anyway) and so came the long neck. Now, the giraffe could eat the leaves on top of the trees and no other animal could. The trade-off was that, if it ever tipped over, it would never get up again. Or be able to hold a glass of Chardonnay.

There have been countless trade-offs. For example, millions of years ago, when the tropical forests disappeared because of shifts in the Earth's crust, the Great Rift Valley in Eastern Africa was created and apes found themselves treeless. Without a jungle, there was no need to swing from branch to branch so *boom!*, the bipedal creature was born. I guess someone thought two legs were better than four. Now that we were hands free, we could make tools and (more importantly) jewellery while being able to stride long distances. We needed to walk, and walk fast, because the violent temperatures on earth forced us to move across the planet without burning our feet.

Another bitch about standing up is that women now have difficulty in giving birth. (Oh, really? I hadn't noticed . . .) On all fours, it was easier to deliver but, standing up, the pelvis is too narrow, so pushing out the baby is more painful than passing a beach ball.

Again, I'm absolving myself from bad personal choices now that I know my brain is making decisions I'm unaware of, some of which are far from beneficial. Part of the problem is that the different regions in the brain aren't all on the same page.

The Good Times: When We Were All in the Same Boat

When we as a human race were young, life was dandy – apart from the threat of being devoured, or frozen by an Ice Age. During our ape epoch, we lived in tribes of about thirty to fifty, most of them family members or, at least, very close acquaintances. Everyone was on a friendly 'Hi' basis. In this environment, we could trust each other because nearly everyone shared the same genes. Of course, the bad news with all this in-breeding was the kids often had webbed feet or only half a head.

These were the good times of the hunter-gatherers, and they lasted many thousands of years. The men did the dirty work, spearing dinner; the women peeled roots and bulbs (this was before Women's Lib). No one complained, mainly because they couldn't speak back then; language hadn't been invented yet. If a man wanted to date a woman, he would raid the next tribe, drag the woman back to the camp by her hair, have sex with her and probably never call her again. These were not particularly romantic times (pre-Valentine's Day).

For you animal fans, thirty to fifty was also the ideal number for a group of chimpanzees. At that number, it was possible for everyone to groom each other for bonding purposes. Everyone had a chance to pick bugs off everyone and no one felt left out. When a chimp population went up to a hundred or more, the social order was ruptured and there was internal squabbling.

We humans, in the meantime, could expand our tribes to up to 150 members and still maintain equilibrium and trust. If someone was in distress, the rest of the tribe would come to their aid with flowers and chiselled-in-stone get-well cards. They could groom by proxy.

A hundred and fifty was ideal then, and still is today for successful family businesses, social networks, town-hall meetings, military troops and harems. Everyone, if not close, was on nodding terms; no one was a stranger, so there was no need for ranks or laws.

We all had a job to do: picking, skinning, hacking (obviously, not the way we hack computers now, but to clear the bush). No one was dissatisfied with their role in life and, if you weren't svelte, rich, savvy or Jennifer Lawrence, you didn't feel like a toad.

No one laughed at you because your teeth jutted straight out of your mouth, because everyone's did. No one suffered from low self-esteem. It hadn't been invented yet.

Ranking and Status: The Birth of Envy

One day, for the sake of competition for food, territory and sexual partners, we started to rate ourselves against the next guy. We suddenly became conscious of who was weakest and who was the most popular. This developed into feelings of shame, low self-worth and self-criticism for those who felt low down on the totem pole of status.

From this point onwards, the idea that we were all equal was axed. Now, to 'make it' in the community, people felt under pressure to bring something special to the tribe that would make them stand out. Anthropologists found physical evidence of this when they dug up the bones of women who lived thousands of years ago. They found private burial sites where women were bedecked in jewellery. The unjewelled were all flung together in a communal grave. Other evidence shows that the stronger and more savvy a man was, the more likely it would be him that was first in the food queue and fed the most. This also went for the women with the widest, child-bearing hips. (Unlike today, where you have to look like a stick with one eyebrow.) So, ranking began because of the need to stand out in a crowd. I'm almost certain that's why comedians evolved and still exist today. If you didn't have bulk or child-bearing hips, you might have been flung to the predators as an appetizer, so certain neurotic people in the tribe started to make funny faces or pretended to slip on banana skins, and everyone laughed. It must have worked because, from then on, the face-pullers and slippers got a piece of the buffalo pie. If you didn't have anything special to bring to the social table, you are probably not alive today. Then as now, social status meant survival. Alpha men, ripe young women, those with high IQs and a few comedians are still on the survival A-list.

Another reason why comedians were suddenly 'in' was because, when the tribes became larger, they used rituals, music and comedy to bond the community. It turns out that music and laughter activate the same endorphins as grooming does. Even among great apes, the replacement for mutual grooming was shared laughter, and they had banana skins by the boatload to use for their shtick. The rapid exhalations you make when something amuses you empties the lungs, leaving you exhausted and gasping for breath. This stress on the chest muscles triggers endorphins, which are contagious. If an ape walked into a tree and fell over, everyone found it hilarious and it made them feel great.

Agriculture

(I know it's not the most alluring of topics.) This marked the end of the good times but the beginning of civilization, and there was another big trade-off.

Towards the end of the last Ice Age, about twelve thousand years ago, populations started to increase and settle into villages. Hunting and gathering were out, and farming was in, meaning people remained stationary to grow their own food, which they could replenish each year. People fenced off their plots of mud to keep invaders out and claimed it as their property (birth of the 'me' and 'mine' concepts). They built houses for safety, to separate themselves from scavengers, and from then on, they became more self-centred creatures. They began to accumulate 'stuff' (furniture, animals, probably jewellery), to which they gave enormous value and defended to the death. Raiding was a problem, and lasted for the next five thousand years. Eventually, an elite group appeared who accumulated more than everyone else because they were the bigger bully, taking the peasants' surplus food and, worse, taxing them on it. At this point, 90 per cent of the population were peasants who worked the land and the remaining 10 per cent lived off them. The 'them' and 'us' society began for real.

Afterwards came improved transportation, enabling more people to form villages, which turned into towns, then cities, then kingdoms. The problem was that humans had evolved for millions of years in small tribes, and the sudden increase in the speed of the growth of these empires during the Agricultural Revolution didn't leave enough time for mass cooperation to evolve alongside. This could be why we're mentally skewed today; we didn't have enough time to adjust to our speed-of-light advancements. Some people think that the Agricultural Revolution put mankind on the road to progress, while others argue it was the road to perdition. If we had carried on cooperating with each other, we'd be much happier and better adjusted today.

History Continues . . .

After this, each epoch in history had to deal with population explosion and the raping and pillaging that went with it. During the Greco–Roman period, people blamed their misfortunes on the gods, from thunderstorms to boiling to death in their Roman baths. If we drank too much, it was because Bacchus, the God of Grapes, made us do it. If you suddenly fell in love with Mr Wrong, it was because Venus had sent one of her naked Cupid boys to zing you with an arrow, so it wasn't your fault.

My husband, Ed, never thinks his actions are his fault either. He endlessly walks into low ceilings and complains that cars drive into him, then blames it on the ceilings and other drivers who can't drive. He won't believe that these are his fault for forgetting to bend down and refusing to believe he can't drive.

Jesus

On the cusp of BC and AD, Christ took over and told us He knew what He was doing and we didn't. We were all scum of the earth and natural-born sinners but, if we admitted our scumminess, we could get into Heaven and He'd forgive us.

Jewish people also feel they've done something wrong but usually believe it is someone else's fault. They have holidays where they celebrate how badly they've been treated. They have Passover because they had to leave Egypt with no warning. They were exiled so quickly their bread didn't have time to rise (but matzo, to my mind, is more delicious than bread, so what's the problem?). At Passover, they believe someone else ripped them off so they eat horseradish to remind them of their bitterness (like they need reminding?).

Much later, with the advent of kings and kingdoms, people didn't question their place on the God-approved hierarchical ladder. The king was the cherry on the pie, his court the whipped cream, then waaaaay below was the rest of society and the bottom crust were the peasants. They knew their lot: to be down in the mud planting some bulbs or baling hay. No peasant suffered from low self-esteem because they knew they were the lowest of the low with no chance of an upgrade. There's a certain satisfaction in knowing you can't go any lower.

Calling Dr Freud

Then Freud came along and said that everything was our fault and we should pay people like him to root out our id or wild-man tendencies. He was the first one to say that our problems came from our deep unconscious, which, to this day, no one can find, but we all know it's there. It's like the Black Hole of Calcutta of our Souls.

We could take control and find peace but only if we paid a lot of money and lay on a sofa for years talking in a stream of consciousness while someone behind us took notes. The problem was we found out that everything *is* our fault. Not Zeus's. Not Jesus's. Not nobody's but ours. The Jews were right, we're all guilty.

And now to the twenty-first century, the century of the self, where it's all about me (well, not just about my me, about your me too). The word 'individualism' took its first curtain call in

America in the sixties, and the culture of narcissism began. Our suffering is directly related to our self-involvement; that sense of never having enough and demanding what we want when we want it, which is *now*.

We believe we're in charge of our destiny. I was told when I was young that I could climb every mountain, ford every stream, follow every rainbow . . . well, you know how the song goes. How wrong that advice was. You can climb and ford as much as you want, you are *not* going to find your dream. If we believe this babble from *The Sound of Music*, it will only lead to heartache . . . and irritable bowel syndrome.

Today, we're at the zenith of the 'each man for himself' age, and this is the source of our discontentment. We're on a constant search for who we can blame for this unhappiness, so what we do now is pick on someone of a different colour, religion or ethnic background. We've lost any sense of community and that was the glue that held us together in the first place. Inclusion is the condition where humans flourish best.

Today's Trade-offs

I don't want to be a party pooper about the success of the human race but, trade-off wise, things are out of kilter. Today, more people die of overeating than of starvation. In 2010, malnutrition and famine killed about 1 million people but obesity killed 3 million. More people will die because there is too much salt in their diet than in any contemporary conflict. More people commit suicide than are killed by soldiers, terrorists and criminals combined. We are now our own biggest threat. We still haven't grasped the fact that there never will be an end to craving for 'stuff' because, no matter how much we acquire, we will always lose it, or it will rust and disintegrate, just like we all will someday. We forget that we are all biodegradable. Wanting more is what's making us sick.

Wanting something is different from chasing a goal. It's that niggling, agitated feeling that we haven't got what we're after. Goals are biologically motivated. No one ran over vast terrains or climbed mountains for fun, they ran and climbed to find food. Now what's with us? Triathlon runners? Give me a break. Explain to me what their purpose is in the modern world when all they do is sit behind a desk for a living? I know a little jog is healthy but rehearsing for Iron Man? Please. You're never going to have to swim the Nile, run over the Rockies and bicycle the length of the Amazon, believe me. This is all filed under 'wanting', as it serves no purpose other than getting enlarged pecs, and that does not affect our survival in any way.

In the past, we lived for spare time and time off. Now, we live to get things done faster and more efficiently. A few decades ago, we'd meander over to an answering or fax machine and respond to the messages in our own time; we didn't have to answer anything if we didn't feel like it. You could send a pigeon back with a note, but if the other guy didn't get it you could always say it got lost or shot. Now, if you don't answer an email within four seconds of it arriving, people think you've dropped them and will probably delete you from their contacts. This yanks our primitive fear-chain of being dumped and made tribeless.

People used to leave the office behind when they went home. Now, thanks to emails and smartphones, we never leave the office. The Japanese have come up with a whole new fatal illness called *karoshi*, death from overwork; our ancestors would spin in their graves if they heard about that. Money used to mean you could only spend what you had in the bank but now, credit cards have flung the spending floodgates open wide. We used to shop until the stores shut, but these days they never close and, if they do, we go online. You don't even have to leave home, you can shop in bed.

Back in our prehistoric past, after a hard day's killing and rooting around, we'd retire to the fire. This habit of sitting around the fire continued until recently and was only replaced in the fifties, when we started watching TV. You would still be together, but talking and reflecting were left out of the equation. Today, we're not even sitting on the same sofa or staring at the same square of glass, we're all lost in our own private screens, unaware that there's even a fire to sit around.

In 2014, in a research study, a group of students was asked to sit quietly and reflect for ten to twenty minutes. They were hooked up to a machine that administered mild electric shocks and were told that if and when they got bored they should press the button and give themselves a shock. One guy did this 120 times over 20 minutes. About two thirds of all the male students and one third of the females pressed the button at least once. They found the experience of sitting and being quiet so unpleasant they preferred a shock to break up the anguish of doing nothing but thinking. The experiment demonstrates how easily we give in to any distraction, however disturbing it might be. It stops us from having to confront any of our thoughts or feelings. This is probably why so many of us choose to do extremely challenging things, just to keep our minds occupied, like buying and building Ikea furniture. Thousands of books tell us how to have a more peaceful, undisturbed life but, in truth, very few people want to have one.

2
Thoughts

Don't let me give the impression that thinking is bad – on the contrary: you wouldn't be able to read this book if you didn't think, let alone find your shoes in the morning. But our problem with thinking is that we can't distinguish between the thoughts that are helpful and those that drive us nuts. The deal is that when we have a tangible challenge, whether it's figuring out the speed of light or finding Hoover bags, someone, somewhere, will come up with a solution. But we will never fathom our emotionally motivated thoughts about why we aren't younger/richer/happier/more gifted. Thinking about these things only leads us down the rabbit hole of rumination and sleepless nights.

'What are thoughts?' I hear you ask. Are they products of this thing we call imagination? Let me be the first to tell you that imagination is a physical reality and not an elusive thought bubble in your brain. Yup, even when you imagine something, a physical and biological occurrence is happening in your brain and body.

As soon as we created language, we could think in words and sentences. The brain uses the same regions for talking out loud as it does when you speak to yourself in your mind. When you sing a song to yourself, you're using the same auditory cortex as if you were hearing it in the external world. The same thing goes for when you picture a scene in your mind: you're using the same visual cortex as when you actually see something.

We like to think that our thoughts are who we are; we imagine our head to be a giant computer. We are the stars in our own reality shows and everyone else is a bit player. Even though we know other people have their own shows and opinions, deep inside, we feel it's because they're deluded and not so bright. This is usually why we smirk and roll our eyes when other people give their opinions. We sometimes forget – everyone is watching a different 'me' channel.

'I think therefore I am,' Descartes wrote, which I always thought was clever of him. But it turns out he was wrong. You are not your thoughts, you're a consolidation of millions of processes, none of which you're aware of. In a single day, we grow 40 trillion new cells and 40 trillion die. Who comes up with these numbers? I've often wondered. While I'm at it, I might make up some too. Okay, in your lifetime, you will have grown 8.2 miles of fingernails and filled 561 ton-barrels of urine. Impressed? And all this is happening while you're choosing a carpet, blissfully unaware of all the comings and goings inside you. Nothing in you ever stops working. There's never a day off until the final curtain so it turns out it's more like 'I am, therefore I think.'

What you are is much bigger than your thoughts. In fact, thoughts make up only about 1 per cent of what's going on inside your brain. The other 99 per cent of the mental caboodle you're unaware of and haven't got the bandwidth to ever know. Your brain is too busy to bother with thoughts because it's having to sort out about 11 million bits of information per second. It's a miracle we get any mental information at all with those statistics. There are a few things we're aware of: for example, we know when we have to go to the bathroom.

I would say that's not incredibly impressive or earth-shattering; squirrels can do that.

Bees

To help you with this concept, picture your thoughts being manufactured by a single queen bee sitting in her larvae in your brain (for those of you who are new to neuroscience, there's not an actual bee in your brain). Around her are waiter, room service, maid, construction worker, valet parking and plumber bees. Okay, now imagine that in your brain there are also bees in charge of your actions and thoughts. Say some of them are watching films of coffee in your visual department; others are manufacturing the smell of coffee in your smell department; and the movement bees are manipulating your feet towards a Starbucks. The queen thinks she wants coffee but she's deluded, it's all the bees working in their separate departments sending in their votes which motivate her to go for the latte. There is no one bee that makes the decision, it's whichever department buzzes the loudest.

So when you say, 'I think I'll have a dry skinny double cappuccino ginger and pumpkin blend picked in Nicaragua by eco-friendly slaves,' you'll get and drink the coffee while some bees are already putting together plans for a chocolate muffin. Even though you/the queen will be (arrogantly) certain that you came up with the plan, there are actually a million bees working ahead of you.

It turns out that, when we do anything, it's not because we had a thought and then acted on it, it's the other way around. If either of these thoughts came into your head, say, 1. *I have to buy this shoe because it's on sale and there will never be another shoe like it at that price* or 2. *I think I'll devote my life to saving Armenian turtles*, you may think they have just popped into your head, but the choice has already been made by the 99 per cent bit of you you're not aware of. Thoughts are just the tip of the iceberg. If you purchase the shoes or air vac a turtle to safety, it's already been decided. So much for free will. Our thoughts are like hitch-hikers to the rest of the brain.

For you neuroscience nuts, here's the scientific explanation of the activity in the brain. For the rest of you, you don't have to read the following if it gets too complicated – just stick with the bee metaphor.

More Advanced Information

You are physiologically equipped with various sensory receptors which receive all incoming information from the environment. If you didn't have these sensors or electrical impulses sending information to your brain, you wouldn't even be aware that there is an external world. The brain lives in total darkness until it gets a message. Go down a manhole and sit there – that's what your brain experiences until it gets some stimuli. The stimulus it receives is nothing more than energy molecules: photons entering our eyes create vision, vibrations through the air enter our ears and create the sounds we hear, and only because certain chemicals land on our tongue do we experience taste. All our experiences are nothing but electrochemical energy, and that's what we call reality. Sorry to reduce you to a telephone switchboard, but that's who you are. Bummer, or what?

When you pick up a signal through your eyes, ears, nose, mouth or, really, any orifice of choice, to enable you to experience any of these senses, various areas in your brain give their input to create the illusion that it's a single sensation, as in taste, smell, touch (see the bees above). Thoughts are not solid things, they're fleeting events, always flowing and changing. Thousands of experiences play in your head, stick around for a second and are then replaced.

This should give some of you folks who worry too much a sense of liberation because, if there's no 'you', there are no problems. You are so much bigger than your narrative. There is no definitive story of you, so don't try to create one. (Mindfulness will teach you this.)

I don't want to namedrop, but even Plato had a hunch about this. He knew that we can't trust our senses and that what we see is more like a shadow on the back of a cave wall. He completely agrees with me . . . even though there's a slight age difference.

What are We Really?

First, let me clear one thing up. The 'you' who you think you are isn't something that's fixed or solid. 'You' are more like a pile of sand pelted by blasting winds which are, metaphorically, zillions of bits of information sculpting and re-sculpting your neural networks. So the 'you' who started reading this sentence will be a slightly different 'you' by the time you finish it. You're shape-shifting millisecond by millisecond. It's memory that keeps your sense of self cohesive, meaning that you don't wake up a feminist one day and the next become a clown.

Hurrah! We don't have to search for meaning any more! Some people dig around to find out who they are, like they're looking for the toy in the Christmas cracker of their psyche. I wouldn't read anything into dreams either. In my opinion, they are just a random grab-bag of sleep detritus. Some people make a profession out of helping their patients piece together bits of rubbish from their dreams, but to me it's like trying to read your future by looking at what you left in the loo. (Forgive me, Freud.)

My Story

I have dreams that would get me hospitalized, if I took them seriously. I once went to a shrink to help me analyse them. I told him I had a dream about a cow who takes me to school on my first day. An elephant shows up, which I find even more embarrassing than the cow. Then the elephant turns into a balloon with a baseball bat and bats a ball into the outfield. Everyone involved boos, especially the cow. Get this, I was told by the shrink, the cow represents my mother, for reasons unknown, but maybe because a cow has udders and so does my mother, in a way. The elephant represents my father, with his dominant, dictatorial ways, which were daunting. My father then turns into a balloon (?) and the ball he hits into the outfield symbolizes me. The shrink told me this suggested that my father didn't hold out much hope for me ever making a home run; anything I try to achieve, my ball will always end up in the bleachers. And *voilà!* I was charged £120 for that evaluation. I felt much better (not).

Fifty Shades of Critical Thinking

Above, I've explained that our internal thinking is a small part of who we are, but it might be of interest to some of you to know why we think mostly negative thoughts. These days, around 80 per cent of our thoughts are negative. But why?

We humans don't use instinct as animals do. We have to figure things out with our brains. True, when we were in our fish phase in the sea, we got the flippers without having to think too much about it, but we didn't stay in the water long (don't ask me why). For some reason, we crawled on to land and, suddenly, we needed legs. Sadly, those legs didn't help us out when running away from our friend the sabre-toothed tiger, so we needed to make spears, then fire, and by this point things got so complicated that, if we were ever going to stay alive, we would need to start thinking. This new thing called thinking meant we could use abstraction, sequencing, prediction, imagination and make decisions. Along with these ritzy new features came worrying about the future, churning up the past and the thing that makes us crazy today: rumination.

So much of our language was built around warning us of dangers (which really existed), and these messages became internalized. So what started off in the past as something helpful, like, 'Oh my God, I'm going to be eaten or caught in another Ice Age, without any gloves,' has become, 'I'm going to get a million dislikes/no followers/rejected on Tinder/lose my job/girlfriend/looks/money/life.' Our inclination is to lean towards the negative, so the very thing that once saved us now drains and disables us, keeping us on a tread-mill of worry.

If we receive good news, we can always think about it tomorrow, whereas bad news has to be dealt with immediately or it's curtains. Bad experiences are fast-tracked to brain central; from this, the concept of 'once bitten, twice shy' was created. Rick Hanson, a very good-looking and smart neuroscientist says, 'The brain is Velcro for negative experiences and Teflon for positive ones.' I know this metaphor in every cell of my being because I am almost 98 per cent Velcro.

My Story

When I was in my twenties, I got into the Royal Shakespeare Company and felt that out-of-body experience with the accompanying stab of joy in my heart. I told everyone I knew that I had got in, although many didn't believe me: I was a terrible actress. For the RSC audition, I did pull off one of the great auditions of all time, presenting a monologue from *Antigone* with full saliva-spitting and deep, sobbing verve. Maybe the people auditioning me thought that this was normal behaviour for the Greeks. I mean, when you find out your brother was eaten by his aunt, what do you expect? (I didn't read the rest of the play, but I know something bad happened.) I saw Trevor Nunn (at the time the artistic director of the RSC) out of the corner of my eye. He had been eating an ice-cream cone when I began and, when I'd finished, his tongue was still out but the ice cream was down his front. Anyway, I got in and was offered the role of a wench in *Love's Labour's Lost*. I was ecstatic.

I remember the feeling, but what I remember even more was being on stage with remarkable actors like Alan Rickman, Zoë Wanamaker, Jonathan Pryce, Richard Griffiths, Michael Hordern and others who were or became famous. All of them could do English accents, mainly because they were English. Mine was very Dick Van Dyke, and I recall the looks I got from my fellow actors when I delivered my lines; they actually winced. I even had a rolled-up note thrown at me during my performance from another actor on stage with me that read, 'You can't act. Get another job.' Sometimes when I spoke my Shakespearean lines, *I* winced at what was coming out of my mouth. When I spoke the Bard's words, it was like a dull, dead drone. I practised for hours in my dressing room – 'La la la la ta ta ta ta ga ga ga, meeeeeeeeem, beeeeeem, leeeeeeeem' – and yet my accent didn't change. That's what I remember, far more than the letter offering me a place and congratulating me on having got into the RSC. To this day, the memory of that rolled-up note is much more poignant. I have always been Velcro for the negative.

We create our own whips with our thinking. What makes us most stressed is not the actual situation but the thoughts that accompany it. As Hamlet says in Act II, scene 2, 'There is nothing either good or bad, but thinking makes it so. To me it is a prison.' Hamlet and I agree on this point.

Inner Theme Tunes

I just want to make a point here. I always think it's so pretentious when people use quotes, especially at dinner parties, to show how intellectual and well-read they are. They'll say something like, 'Speaking of fish soup, I remember Marcus Aurelius said, "Blah blah blah . . ."' Coincidentally, I am one of those people, so here we go with another quote, and there will be many more. Marcus Aurelius wrote, 'Everything we hear is an opinion, not a fact. Everything we see is a perspective, not the truth.' How come nobody knows this stuff except me, Marcus, Hamlet, Buddha and Plato?

The Point of All This

I don't know about you, but for me the realization that thoughts – and particularly the negative ones – are just another by-product of our evolutionary survival kit helps me accept myself, mental warts and all.

What a liberation it is to find out my negative thoughts aren't my fault. Free at last!! They're pre-recorded from Mommy, Daddy, genes, evolution and experience. Negative thinking comes with the package, just as your legs do when they won't run as fast as you want them to or can't take you over a hurdle and you don't start berating them, do you?

The Monk, the Neuroscientist and Me

As promised in the Preface, I'm going to welcome Thubten (Buddhist monk) and Ash (neuroscientist) to explain how the mind and the brain interconnect. Ash can explain both the brilliance and the shortcomings of the brain and Thubten can tell us what to do about it. So, take it away, boys . . .

Ruby: Here's an easy one, Ash, why do we have thoughts in the first place? What are they?

Neuroscientist: It sounds like such a simple question! I think what you mean by thoughts is the internal voice, the way you talk to yourself about your own ideas, who you are and what you want. That's right at the centre of what we mean by 'consciousness'. But we get into trouble with thoughts about thoughts – rumination.

Ruby: And what does Thubten think of thoughts?

Monk: Well, Buddhism is traditionally called the 'science of mind'. It's all about the study of the mind itself. We're looking into the nature of thoughts and the mind which is thinking those thoughts.

Ruby: So, Thubten, you think about the mind and Ash thinks about the brain. I heard that the brain is the piece of meat and the mind is the information that flows through it. Do you guys ever cross paths?

Neuroscientist: There's a bit of a love affair between neuroscientists and Buddhists. Buddhists have been studying mental function in detail for a very long time, and neuroscientists get pretty excited about that. The questions are the same but the methods are different.

Monk: Exactly. It's an exciting interface.

Ruby: Are you guys flirting? Okay, here's the million-dollar question: where are thoughts in our brain? Where do they reside?

Neuroscientist: 'Where' can be misleading in the brain because we're almost always talking about large networks rather than specific areas. So when we look at thoughts that we're aware of, when we speak to ourselves with an internal voice, we're linking parts of the brain that do language, like auditory cortex, with parts that do self-awareness, like –

Ruby: Don't get too fancy, I want people to read this book.

Neuroscientist: But you asked me the question! I'm a neuroscientist, the answer is going to be complicated. Hearing your own thoughts requires both language and self-awareness, so you're talking about the dominant temporal lobe and the medial frontal cortex, mostly the cingulate cortex. That network creates a sense of self, so you recognize that the thoughts are your own. If the network doesn't work, like in schizophrenia, you hear the voices but can't recognize them as your own. The voices will seem to come from someone else.

Ruby: I love when you talk brain to me. Thubten, what's your view on all this?

Monk: We can study the nature of thoughts through practising meditation or mindfulness. Thoughts don't exist in the way we normally think they do, so that blows the question of 'where' out of the water. The whole point is to see that we're bigger than our thoughts – so we don't have to be so glued to them, or controlled by them. There's an old Tibetan saying: 'When you run after your thoughts, you're like a dog chasing a stick. Instead, be like a lion: turn around and face the thrower. You only throw a stick at a lion once.'

Ruby: Ash, I wrote earlier in this chapter about how we're the worst critics of ourselves. What is the point of having thoughts that make us so miserable? At least I'm grateful you can't hear what I'm thinking.

Neuroscientist: How do you know I can't? I'm a doctor, after all.

Ruby: I'm pretty sure you can't hear channel 'me' because you're on channel 'you'. If you ever heard what I'm thinking, you'd run for the hills. There's a lot of self-flagellating going on because I know you're a neuroscientist and I'm not. I'm scared you'll catch on that I know very little. Those are mine; what are your thoughts?

Neuroscientist: I worry that what I'm telling you isn't funny and I'm going to come across as dull and boring.

Ruby: What about you, Thubten? Do you have critical thoughts, even though you're a maroon-wearing monk?

Monk: Yes, I'm worried that if I don't come up with the goods, you'll dump me for the Dalai Lama.

Ruby: I was thinking of doing that but he's not free. What's with us? I still want to know why are we so self-critical. Ash, do we have some sort of asshole gene? Nothing to do with the bum. Just asking.

Neuroscientist: If we're using the word 'gene' as shorthand for some biological tendency that we're born with, then I think yes. Our brains have a tendency to focus on error signals; to focus on the negative. That may be something we've developed to survive.

Ruby: Well, where is the asshole gene exactly?

Neuroscientist: The asshole gene is the self-critical bits of the brain, the parts that do internal monitoring. Those are regions like the prefrontal cortex, the anterior cingulate and the insula.

Ruby: But why do we need those parts? Life is tough enough.

Neuroscientist: Because those parts activate error signals, which are there to help you. So, for example, you're walking down the stairs and you think you've come to the bottom step but your foot continues to go down a bit too far. Your brain generates a big error signal that grabs your attention and stops you falling. Error signals work like stop signs, so you can figure out what's wrong and do something different.

Ruby: I can understand there's a part in your brain giving you, 'Oops, maybe I'm going to fall down the stairs' but why do I also add, 'I'm such a klutz?'

Monk: The problem is we have that error signal and, on top of that, judgement and self-criticism. This seems to be especially prevalent in modern society. When the Tibetan lamas first started coming to the West in the sixties, they were shocked at the levels of self-hatred and guilt that so many people here suffer from. In the Tibetan language, there's no word for 'guilt'. In fact, in Buddhist cultures, children are brought up to believe that the mind is naturally good.

Ruby: What if, in the West, we left out the word 'guilt' and didn't teach it to kids? Would we not feel guilt?

Neuroscientist: We'd definitely be better off. Our brains may stop and notice our mistakes, but if the culture doesn't push it then guilt may never come into play.

Ruby: If we gave monks the word 'guilt', would they feel it after a while?

Monk: A lot has to do with upbringing. I think children in the West are told 'no' a lot of the time. Maybe our parents are so stressed that it's just easier to say 'no' a lot. Traditional society in the East tends to be more of a 'yes' culture. If children hear the word 'no' often enough, they grow up believing they've done something bad, haunted by this internal voice, 'You're wrong' or 'You're bad,' and that who they are is something to be said 'no' to. We've grown up with the feeling that there's something wrong with us.

Neuroscientist: I had a lot of guilt as a child. It's probably the intersection of an American Midwestern culture, which is pretty puritanical, with an Indian culture where children carry the burden of their parents' aspirations. I suppose that, as an adult, that guilt is at the root of a lot of my self-criticism about something being wrong with me or whether I'm working hard enough. But can I go back to the asshole gene for a minute?

Ruby: Maybe because you feel guilty you want to change the subject. Okay, go back to it.

Neuroscientist: I think in some ways it's not bad to be self-critical, because it warns you that something isn't right. Your internal voice grabs your attention so that you analyse the situation. When you're feeling happy and things are going well, the internal voices are quiet.

Ruby: Why don't you notice the nice voices? You're so right. I've never gone, 'Wow, I am so gorgeous, I could date myself.'

Neuroscientist: When you're feeling happy, you're not doing a lot of analysis about how you're feeling, you're just feeling. But negative thoughts are different. They signal to the brain that there is a problem and resources should be dedicated to fixing it.

Ruby: So, if I had a life with no negative thoughts, what would I be thinking about?

Neuroscientist: I think you'd just be being. When the brain isn't worrying and you're just at rest, it's neither happy nor sad, it's just whirring away. Thubten, is that what mindfulness is?

Monk: Mindfulness isn't just the absence of negative thoughts, it's about finding a completely different relationship to thoughts, negative or positive. Whatever you're thinking, the aim is to just notice the thoughts without judging them, and then they don't have so much power over us. The point of learning mindfulness is to reduce our suffering; and the reason we suffer is because we believe our thoughts and therefore we're constantly being pulled into distressing mind states.

Ruby: So, are you saying we're all natural-born sufferers?

Monk: We tend to be what I call 'nego-centric', made worse because the culture we live in constantly makes us feel we're 'lacking' – we're not beautiful enough, thin enough or driving the right car. There are various ways in which we suffer. We could talk about four major types of stress: not getting what we want, getting what we don't want, protecting what we have, losing what we're attached to.

Ruby: That's me. All four of those.

Monk: And they just continue. The wanting creates more wanting, and it creates a feeling of deficiency or lack.

Ruby: So, what do you do about it?

Monk: It takes training. That's what mindfulness is for. You see it arise, learn not to grab on to it and, because you don't make a meal of it, it can begin to dissolve.

Ruby: I guess we're born with the ability to stand back from our thoughts and not give them bad reviews, but we forget to do it. Same as we're all born with a pelvic floor but we don't use it. Do you agree, Thubten?

Monk: I don't know what a pelvic floor is.

Ruby: I'm not going there, I don't want to ruin your life. Ash, do you think you can hold back from judging?

Neuroscientist: I'm very judgemental. In general, judgement and evaluation are central to how the brain works. But there is a gap in time between perceiving something and judging what it is and what it means. For example, when you see an object, it might take something in the range of fifty to seventy-five milliseconds to start to see its colours and shapes. It takes up to two hundred milliseconds before your brain can name the object and associate that name with meaning and value.

Ruby: I heard it was about 201 milliseconds.

Monk: I heard it was 199.

Neuroscientist: I'm pretty sure it's two hundred. Maybe the goal of mindfulness is to intervene after those first seventy-five milliseconds, to just see what's in front of you before naming it or evaluating it. I imagine that could be possible with training and practice.

Monk: And it's the same with thoughts. Mindfulness trains you to pause in those early milliseconds when you notice a thought, before you start evaluating and judging. Then the thought will control you less, and you can start to make wiser choices.

Ruby: So, I'd just notice I missed the stair but I wouldn't need to then tell myself I'm an idiot for missing it.

Neuroscientist: Right. You can develop a habit like that. You have the thoughts but you don't get wrapped up in the commentary.

Ruby: So, what part of Thubten's brain stops him from getting caught up in judging and commenting? If we skinned Thubten's head, what would we see?

Monk: I'm already skinned, I'm bald.

Ruby: Not enough. I want to know what his brain looks like after all that mindfulness?

Neuroscientist: Well, studies show that the right dorsolateral prefrontal cortex becomes more active in experienced meditators. That part of the brain is involved in stopping self-criticism and judgement. The more you practise taking control of your thoughts, the more effective that part of your brain gets.

Ruby: So, the idea is to bulk up that part of the brain so we can pick and choose our thoughts, like you do with Spotify?

Monk: Yes, with mindfulness we can start to understand which thoughts are helpful and which ones make us suffer, and we strengthen our ability to let go. We're not usually our own boss; our mind often goes to places we'd rather it didn't, or it won't do what we want it to. But now we're getting into the driving seat.

Neuroscientist: That makes sense from a brain perspective, but how do you actually do that? What are you doing when you're practising meditation?

Monk: When you notice your mind getting caught in a whirlwind of thoughts, you can gently bring your focus to one of your senses, such as breathing or sensations in the body, using these as an anchor. It's like getting in the driving seat and driving that car back to base.

Neuroscientist: Okay, but how does that help stop your runaway thoughts?

Monk: Because you can't be focused on thoughts and breathing or sensing at the same time. And when you get good at this, your thoughts are going to control you less. Eventually, you won't need an anchor to bring you back, you can just observe and let the mind be without needing to interfere in it.

You'll find the relevant mindfulness exercises for thoughts in Chapter 11.

3

Emotions

Gloria Steinem wrote, 'The truth will set you free. But first it has to piss you off.'

What's the Matter with Emotions?

If it wasn't for emotions, we'd never need shrinks or meds. Sometimes they're what make us suffer, even more than physical pain, which is why we spend a lot of our waking hours trying to bury or run from them. They rise up from within and we're held hostage by them until they decide to slink away. We can build telescopes that allow us to see a star 345,678,803,940 light years away (I'm not an expert in this area, so the numbers may be slightly wrong), but do we have any control if we fall in love with someone twenty years younger than us with the brain of a fruit fly? No, we do not.

The fact we can't figure out how to be happy feels, to many of us, like a personal failure. If you don't believe me, go to the self-help section in any bookshop. If you lined up all the books on how to be happy, they would circle the equator fifty-seven times. You're even supposed to feel happy on your birthday, which is the most miserable day in my calendar, along with (Happy?) New Year. Happy about what? All of us whizzing faster towards extinction? What makes me deeply unhappy is the feeling that everyone else has figured it out. This feeling of helplessness creates a kind of background buzz of discontentment, along with a slash of envy.

Thank God: Medical Help

If emotions do become overwhelming, they can be dealt with because, luckily, some biologist in the second half of the twentieth century was looking for a cure for tuberculosis and accidentally tripped upon the recipe for antidepressants. Around the same time, some guy in Australia found that the lithium he was feeding guinea pigs made them docile (why he was giving them lithium I do not know) and, accidentally, he discovered bipolar-disorder medication. To this day, there are very few manic-depressive guinea pigs, which is a good thing; they were getting too crazy, zipping around on their running wheels for months on end, buying thousands of pounds worth of food pellets and selling them the next day. Soon afterwards, scientists found that tiny molecules transported to the brain could banish anxiety. Hello, Valium! Welcome, Xanax! These are still very popular emotion squasher-downers.

A History of Emotions

We've had emotions for around 100 million years while, in the timeline of our evolution, we've had language for maybe ten minutes. I'm always so intrigued with these timelines some crazy anthropologist created to give us an idea how long we've been in existence, based on the twenty-four-hour clock. Usually, it runs along the lines of 'If the Big Bang happened at midnight, then at about 11.40 a.m. dinosaurs appeared. At 11.41 they were gone. At 11.56 apes showed up.

At 11.58 humans made their entrance and, for the next few minutes, there was a profusion of fashion wars as styles came and went, came and went (buffalo skins, codpieces, fans, pointy wimple hats, wigs for men, the mini, Marks and Spencer's underpants) and, at a nanosecond to midnight, Donald Trump became President of the United States and that just about finished off existence as we know it. Now, we're just waiting for the black hole to suck us back in and start again . . . but, this time, with no mistakes.'

The Point of Emotions

Many of us don't realize that emotions were originally created to ensure our survival; they meant well. We first pick up their scent in our bodies, created by various cocktails of chemicals that provide us with moment-to-moment feedback on what to avoid and what to approach. To avoid danger, a weasel sniffs the wind, a snail ups its antennae, an octopus extends a tentacle, while we humans use our emotions to test the waters. Those feelings tell us what's safe and what's unsafe, but do we always listen to them? No, we do not.

We feel the emotion first and then, about 200 milliseconds later, it's translated into thoughts. The reason we feel something faster than we think it is that if there is an emergency and we have to wait around to think the words 'I should get out of here!!!' we'd be toast. So, it's emotions that save us during emergencies, not the thoughts, and it's emotions that signal to us how we feel even before we think and label the feeling. When we feel a vice-like grip in our bowels, there's something to fear, and if we go soft and gooey around the heart area, we're probably watching a Disney film and some woodland creature with big eyes just died.

Our thoughts can't really express what we physically

experience as emotions. Our vocabulary is limited; words are a trickle of water compared to the Niagara Falls of emotions. There are thousands of emotions but we can only verbally translate a few dozen. Our language, unless we're a poet, is totally inadequate to express all our emotions. On the other hand, we need language to be able to talk about our feelings; if we didn't, we'd be emotionally constipated. We need to vocalize our emotions, otherwise, one day, they will erupt out of us like Vesuvius. Suddenly, out of nowhere, one Christmas morning, you'll try to beat your mother-in-law over the head with a plunger and you won't even know why.

These days, discussing our feelings, especially if they're heartfelt, is looked upon as indulgent and queasy-making. But an emotion is not some girly, hormonally imbalanced flash in the pan, emotions are concocted by complex neurochemical systems which give us a sense of our physical selves. They are what give us a sense of our connection with the world and with one another. Other mammals feel love for their kin and the pain of separation but we humans are notches above them because we can think about these feelings and weave them into literature, art and mind-numbingly 'blah' songs on *X Factor*. The only people excused for having no emotions are the walking dead, the CEOs of some big organizations and those with specific brain disorders. Otherwise, you have no excuse. You can be blind, deaf, missing any number of limbs or toes but, when you lack emotions, in a sense, you're no longer human.

There's a famous story about a guy, Phineas Gage, who was working on the railroads and accidentally drove a metal rod through his skull (God knows what he was doing when that happened). After the accident, he was fine cognitively – he could speak, think, remember things, but when he saw his family and friends he had absolutely no feelings towards them. He knew who they were but felt nothing. Everyone around him suffered but he said he was perfectly happy, which just shows that it might be worth hammering a rod into your head if you're sick of your family.

What are Emotions for, Especially the Bad Ones?

When it comes to human development, nothing is an accident, so even negative emotions had some reason to emerge; otherwise, they never would have appeared in our evolutionary grab-bag. Everything we've got going for us, physically and mentally, was created for a purpose. Some of these emotions may be a burden now, giving us ulcers and acid reflux, but they're what kept us alive (see evolutionary trade-offs).

Fear

If we hadn't felt fear in the past, we would probably have been skinned, de-boned and eaten.

Rage

We needed rage to frighten off the enemy. We didn't turn it on ourselves back then.

Anxiety

Anxiety ensured that we were prepared for attack. It motivates us to remember what we did in the past in similar situations to ensure our future.

Disgust

Disgust was a necessity to spot which foods were poisonous and warn others not to eat them by wrinkling our noses and curling back our lips.

Shame

In the past, the tribe was everything to us and our acceptance into it meant life or death. If we sensed that we let other members of the tribe down, we felt shame. That horrible kick in the stomach motivated us to do better and work harder. This was a healthy shame, which promoted working for the benefit of the group. The unhealthy shame that we have today, one where we feel we're not attractive enough, or something equally banal, is different and serves no purpose to help our tribe. Please tell me how the hell being 'pretty' benefits the gang? Now, we feel shame because someone rejected us on Tinder or didn't give us a thumbs-up for the photo of our lunch that we posted online.

Our obsession with the self, rather than thinking about the success of the group, is exactly the problem with the human race today. When it became more about 'me' than 'we', we lost the ropes. I wonder who first experienced this narcissistic sense of shame? Maybe, one day in the past, a caveman looked at his etching on the cave walls and thought, *Well, that sucks.* Maybe there's a fossil somewhere of a cave woman looking at her bum and wondering if it's too big?

Animals can feel something akin to shame when their status is threatened. They, however, don't think of themselves as losers. They don't give a toss what other people think, they can even pee in front of everyone in the middle of a party.

Guilt

The origins of guilt are a different kettle of guilty fish altogether. The difference between guilt and shame is that, with guilt, you feel it not because you feel inferior or a weirdo but because it drives you to fix the situation; to want to make amends. Shame comes with self-disgust. Again, animals don't have these feelings, they just do their thing. This is probably why there aren't many Jewish or Catholic geckos.

Grief

This is an emotional reaction to loss, and it was always so, right from our beginnings when we swung from trees. Animals feel grief but don't brood about it; it decreases after a while and they move on. We, with our big brains, can become consumed with memories, the 'what-if?'s and 'why?'s. And, because we keep reigniting these memories, the grief is never allowed to run its natural course. There are rituals that have developed over time in various cultures to allow people to mourn together, which helps the individual to be able to bear the pain. When someone gets exhausted from ululating or weeping, someone else in the group can take over. In our culture, many of us don't have these rituals, so we have to find our lonely, individual way to deal with grief and, for some reason, we're ashamed to express sadness in public. What we all need to do is learn from those wakes they hold in Ireland for their dead. Everyone's so drunk, they don't even remember that someone's died. But they're together and still have a sense of community, God bless 'em.

Love

We needed love to bond with our young, our mate, our friends and our community to make sure that, every year, we'd get a birthday card.

What Happens in Your Body When You Have Emotions?

I'll give you a refresher in case you happen not to have read my last book. (Why not?)

When we get a scary vibe (which you'll recognize because your hair is standing on end, you've broken out in goose pimples and your heart is pounding), it means you're pumped to the max, ready to scram, kick ass or just stand there like a frozen statue. If you stay in that state, the first thing to go down will be your memory, then your immune, digestive and reproductive systems. At that point, what with the missing memories, you won't even remember what your options are.

This is all happening under your radar so you won't be aware that your system is deteriorating or why your brain cells are beginning to atrophy. Trust me on this, we all have myelin sheaths that cover each of your nerve cells (neurons) to speed up their signals to each other. If those sheaths get damaged, the neurons connecting different regions of the brain get weaker and the result is you can no longer put your thoughts together and your ability to be rational goes AWOL. In effect, you've been dumbed down.

If we can't think straight or be rational, we begin to feel threatened, even if there's nothing nearby that can harm us. We start to blame other people for making us feel paranoid, and so begins the 'them' and 'us' syndrome. We stop thinking of 'them' as fellow humans. Another result of neuronal atrophy is that our thinking becomes narrow and rigid and we begin to think that anyone different from us is the enemy. We all have specific fear triggers embedded in our memories which we react to emotionally without knowing why, especially when we're stressed. We're at the mercy of old associations.

If we met just one person with a beard who may have scared us when we were young, later in life we might fear the entire bearded race. This may include Muslims and people dressed as Santa. We can develop a bias even if we just read about a bad guy with a beard. (I don't think I need to drive this one home, but when I see a man with a short little moustache under his nose and a hairdo with a side parting, I run for the attic . . .) We can learn our way out of this with mindfulness, or by growing a beard and joining the 'them' group.

Memory and Emotion

So, none of us is aware why we react the way we do to specific things, people and incidents. We never see anything for what it truly is, only through the interpretation of our own memories.

For example, when we see a cigar, we don't all think it's a penis, as Freud suggested. I'm sorry, Sigmund, but when I see a penis I don't think 'cigar', or vice versa. (The first man I saw naked was John Lennon on the front of his album cover with Yoko, who really was wall-to-wall hair, like a carpet. Since then, when I see a genital, I think of the Beatles. We see what we see because of early associations.)

Gut Feelings

People sometimes say they have a gut feeling about something and then act on it, as if their gut is some wise sage from Tibet. Sometimes a gut instinct is right on the button, but so often it isn't. If gut instincts were always right, there would be far more winners in Vegas than there are. The gut has 500 million neurons compared to our brain, which has

100 billion, so the brain is a little more in charge. Every experience we have is registered at Grand Central Memory Station and every feeling must pass along those train tracks. So, unless you're conscious of that memory, you won't know how you got the gut feeling. It's the awareness of a memory that helps us make better decisions.

My Story

A while ago, while browsing in Selfridges, I was suddenly caught in a full-frontal panic attack and had to flee from the shop to go outside and hyperventilate into a bag. On reflection, I remembered that when I was about eight or fourteen (one of those) I was bitten by a Dalmatian, and that's why I got hysterical: I was trying on spotted leggings. *Aha!* I thought, *I've figured out my fear*, so I went back into the shop and shouted, 'Wrap up those Dalmatian pants. I'm not afraid any more.'

We're Nuclear Giants but Ethical Idiots

In my opinion, in order for us to survive in the future, we have to upgrade our minds in the same way that we keep upgrading technology. We need to consciously develop our emotional intelligence. If we keep lashing out our emotions without any remorse, we might end up a criminal, a rock star or a comedian.

We need to learn how to repackage those emotions and to activate our kinder sides. If we don't get a grip on them, they'll destroy us and everything around us. I sometimes think we take out some of our rage on the environment, and that its demise reflects our fury.

Success should be measured not by our cognitive accomplishments but by the level of our emotional intelligence. That's what gives us the ability to be aware, to self-regulate, to control our impulses and empathize with others. This ability involves developing a stronger prefrontal cortex, which, fortunately, you can grow just like a houseplant (see mindfulness).

The Monk, the Neuroscientist and Me

Over to the experts . . .

Ruby: So, what are emotions? That's the million-dollar question.

Neuroscientist: Do I get a million dollars if I answer it? I guess emotions are really a bodily side of thinking. We can almost always point to a place in our body where we feel we're holding an emotion – love is in the chest, fear is in the gut, anger in the shoulders. We think of those phrases as metaphors, but emotions are how the brain and the body communicate.

Ruby: If I wanted to locate an emotion, where in the brain would I be looking? I always like to know where everything is . . . like on a Google map.

Neuroscientist: The first place you'd look is in the limbic system. That includes brain areas that control hormones, like the hypothalamus, structures that control memory, like the hippocampus, and arousal areas, like the cingulate cortex. From there, you can add the amygdala for fear, the nucleus accumbens for reward and the orbitofrontal cortex for behavioural inhibition.

Ruby: I don't know what you're talking about. Please, Thubten, make it simple for me.

Monk: Okay. Emotions are just thoughts with bells on. Why do we have them? Sometimes they let us know who we are and what makes us tick, but at other times they make us suffer when they're negative and we don't know how to let go. Just as with thoughts, we can learn to observe emotions, both positive and negative, without latching on to them or thinking they're solid or real.

Ruby: But they must have some physical signature in the body for us to translate. When we feel a pang in our chest, how do we know if it's love or acid reflux (to me, they feel the same). I wasn't sure when I met Ed if he was the one or I had eaten a bad pickle. A stab of fear in the abdomen can also be confusing – worry or wind? Call the police or find a toilet? So, physically, is there something going on differently when you have heartburn or heartbreak?

Neuroscientist: There's definitely an overlap. Emotional pain activates the same centres in the brain as physical pain. When you have an emotional sense of suffering, your brain treats that the same as a bodily injury. When your feelings are hurt and you feel a stab in your stomach, your brain is reacting the same way as if you were actually physically stabbed.

Ruby: So, what about happiness?

Neuroscientist: Well, the body and brain respond a lot to negative events but they respond less to happiness. It's too bad, because when you're happy you tend not to notice, but when you're injured you can't think about anything else.

Monk: So, this suggests that happiness would optimally be our default state. When we're unhappy, we feel that error signal. Maybe emotions are signalling to us that we're off balance? And we tend to feel sensations in specific locations when we're upset, but happiness isn't that easy to locate – it feels more generalized, as if it's natural.

Ruby: I can locate happiness. It feels like my chest is drinking champagne. It's throwing up a lot of bubbles.

Neuroscientist: What, do you mean, your breasts are bubbling?

Ruby: Why do you always go to the lowest common denominator, the naughty-boy stuff? The more they're educated, the smuttier their minds are. What do you think, Thubten, when I say 'chest'?

Monk: I'm still thinking about the bad pickle that Ed gave you.

Ruby: Thubten, come back to us. What do you think happiness is? Don't say a bad pickle.

Monk: Maybe we could redefine our idea of happiness. Many people think of it as a buzz, which you get from some kind of trigger. But that kind of dependency leads to grasping. Grasping always makes us feel that things are never enough, and then we get disappointment. Instead, I would aim for a stable state of inner contentment which doesn't require a trigger, something which lasts and is constant.

Ruby: Can't we stop the grasping?

Neuroscientist: Don't we reward grasping, as a society? It shows that you're ambitious, a go-getter. Rewarding a habit makes it hard to break. We have this myth that high achievement leads to happiness because we just ignore all the times that it doesn't.

Monk: The reason people want to achieve success is to find happiness, so why not just cut out the middleman and be happy?

Neuroscientist: Well, achievement does lead to progress, but frantic achievement can come at the cost of burn-out. It's hard to know if it's worth it. When you look at an American pro-footballer who drives himself incredibly hard, when he's young, he's the king of the world, he's the high-school hero. We worship these guys. There was a big spread in the *New York Times* magazine about this recently, and I almost wet myself when I saw it. Neurologists have been talking about this for a long time but we felt like no one was listening. The *Times* printed page after page of photographs of brain sections from National Football League players – that's the top professional league in America. Out of 111 brains that they examined, 110 showed significant evidence of brain injury due to repeated concussion. Those guys hit each other with the same force as running into a brick wall at thirty miles an hour. Their brain injuries resulted in personality changes, depression, divorce and, in some cases, even suicide. There are a few years of stratospheric success, which most of us assume would bring them happiness. But, honestly, is it worth the decades of suffering that follow? How much should we pay for success?

Monk: That's what I'm saying: that high comes at great cost. But with mindfulness you can find real, stable contentment.

Ruby: Someone's going to read this book and think, *Why is the monk always going on about mindfulness?*

Neuroscientist: He does seem a bit obsessed . . .

Ruby: And he's always at it. Sometimes when I'm coming down the stairs, he's sitting there with his eyes shut, and I trip over him thinking he's a fire hydrant.

Ash, if you had had a choice of being a neuroscientist or understanding mindfulness, which one would you have gone for? Would you have the inner smarts or inner peace?

Neuroscientist: I think I'd choose the inner peace. But it's not easy, I grew up with the high-achievement thing too. The first time Thubten and I did mindfulness, I found the whole thing really difficult. I was doing okay but then, for some reason, I suddenly felt depressed. I don't know what about. Then my back started to hurt and I felt desperate to move – actually, desperate to stop doing the mindfulness exercise. Then I started to feel like a loser because Thubten could sit there, and then I felt like a loser about being a loser.

Afterwards, I asked Thubten about that and he said that those overwhelming feelings were just thoughts. That was a big light-bulb moment for me, that my random little thought that I forgot to buy stamps wasn't any different from my sadness that I haven't won a Nobel Prize. For me, it was profound to realize that both are just brain activity.

Monk: Exactly, just thoughts with bells on.

Ruby: Okay, Thubten, I wanted to ask you about all those difficult emotions like anger or fear that we try to avoid. Have you ever been depressed?

Monk: Yes. I got depressed in my retreat. It was very serious and lasted for half the retreat.

Ruby: How long was that?

Monk: It was a four-year retreat. I remember crying a lot and feeling like a failure for being depressed. At times, it felt like a knife was twisting in my heart – it was physically painful. I felt suicidal and almost gave up the retreat. But part of me wanted to fix this thing, and there was nowhere I could go except to learn to engage properly with the meditations. I could leave, but I didn't want to. Things clicked when I discovered how to drop the storyline and relate to the feeling in a compassionate way, just as you would sit with a frightened friend. The pain in my chest started to change and it turned into a kind of joyful feeling.

Ruby: Do you always have that joyful feeling these days?

Monk: I know how to access it.

Neuroscientist: It's interesting that when you stopped pushing the emotions away, the physical pain lifted too. That's a real resolution. We can push emotions to one side, but the body will remember.

Ruby: Where are the emotions in your body? Like, does your arm get depressed? Or your left foot has a bad-hair day?

Neuroscientist: Emotions aren't in a particular place, they come out of the communication between your brain and body. And that's a two-way street, so if you're depressed, even your muscles work differently from when you're in a good mood. For example, posture really affects emotions. Jonathan Miller, who is both a neurologist and a theatre director, tells a story about working with a talented soprano on an opera. Jonathan had a very precise kind of sadness that he wanted the singer to produce, but he couldn't describe it to her. Instead, he slumped into his chair in a certain way, with his limbs dangling off the edges, and he asked the singer to copy his body posture. She was instantly able to sing exactly the sort of sadness he had in mind. Her body understood something that words couldn't convey. I think that shows how emotion is very much a thing that the body does – it feeds information to the brain.

Ruby: Your body makes a phone call to the brain and says, 'Hey, I'm miserable down here?'

Neuroscientist: Yes, that's about it.

Monk: If you're not pushing the feeling away, you're allowing the body and brain to have that conversation without you getting in the way. Then there's the possibility for transformation.

Ruby: I love the saying, 'If you run away, the monster chases you but if you turn and face it, it runs away.'

You'll find the relevant mindfulness exercises for emotions in Chapter 11.

4

The Body

My relationship with my body has not been a good one. I acknowledge it only on a casual basis and think of it as a shopping trolley that carries around my head. My body works wonders for breeding and excreting but, otherwise, I feel it lets me down, especially when I'm among other bodies. In a yoga class, for instance, if I'm surrounded by seventeen-year-olds, I'll beat myself up because I can't get into the extreme Vernashakaka Nozrama Vinhasma pose that they achieve without a flicker of pain. I draw the line at going into that oven and doing what they call Bikram yoga. (Mr Bikram must be laughing at all these Caucasians frying themselves.) Try as I might, I just can't get my ankle round my neck and into my mouth while standing on my head.

My Story

No, my body and I have never got along. When I was a teen, I declared war on my breasts, which were microscopic. Now they're far too big. Why are they doing this to me? And I don't think I'm the only one who's not happy with the way things are down south. Most of my friends complain about their thighs, behinds, ankles, feet and genitals. No body part goes without scrutiny; every cellulite deposit is noted.

Brain/Body

At the same time as learning mindfulness, I've become aware deep down in my now large breasts that your body can teach you as much as your mind, maybe more. Your brain and body are one and the same; every thought, emotion and action is a two-way feedback loop from brain to body, and vice versa. If you change your thoughts, feelings and actions, your body changes. Neurons in the body (yes, they're not just in your noodle) detect the movements in joints, muscles and bones and especially in the 'big boy', the spinal cord, which feeds back to the brain what's going on everywhere in the body like a spy. The brain can then coordinate movement and navigate you through space. (Not Brian Cox space but normal earthly space.)

If you pay attention to the sensations in your body, you get a full readout of the emotional state you're in. When, for example, you're frightened, your body will tell you and if you lower your periscope and look in, you'll probably be tight in the shoulders, your stomach will be clenched and your heart and head pounding (the fight-or-flight response). Your body is making decisions way before you even think about them. It knows how to deal with situations even before your thoughts translate them.

I used to think that emotion was just another aspect of the mind, like language or hearing, but it turns out that it's a combination of body and brain. If you make a pancake out of eggs, flour and milk, you don't ask which ingredient is the pancake; it's the combination. We tend to separate the ingredients but, in reality, the whole organism experiences an emotion.

If you're attracted to or like someone, check what your body's doing. You're probably leaning towards them, mimicking their moves, smiling, and your pupils have become enlarged. This tells the other person you like them so, if you think you're trying to be subtle, forget it.

However when you're mindful in your body, you can recognize the instant you're on autopilot. When I tune in, I usually notice that I'm walking at top speed, even if I'm just taking a stroll with no destination. I live my life as if I'm in a race with myself. Probably the reason I scuttle is because I'm distracting myself from the chaos in my brain. The problem is, you can't out-scuttle your mind.

Ed, on the other hand, is completely unaware when he's gone from manual to autopilot while he's eating. He wolfs down his food, totally oblivious that he's doing it. When I point it out and ask him why, he tells me it's because they didn't give him enough food at public school. I remind him he's not there any more but he just keeps wolfing.

Body Learning

I'm going to use the expression 'body learning' for when you let your body give you an internal weather report. Don't think it's always about slowing down and 'vegging out'. If you need to hit a deadline or are about to take an exam and your body feels like heavy glue, you can choose to wake it up. (Then you can jump, run, take a cold shower.)

If you're furious with someone, and about to let the reptile rip, you know the expression 'bite your tongue'? It exists for a reason. Notice when you feel that physical impulse to pounce. This isn't your imagination, the impulse originates in your motor cortex. The noticing is your pause button. You're giving yourself a few crucial seconds to think about your reaction and find a more suitable alternative to going berserk. If you stand back and notice the impulse, you have time, during the pause, to make a choice, to decide either to attack or to send focus to a bodily sense or your breath and metaphorically bite your tongue. This is ideal as a strategy when you're about to push 'send' on your ranting email. Just like the man on the Tube tells you, you need to 'mind the gap'.

More Body Learning

On the physical side of things, if you practise becoming attuned to how your body feels in each area, you'll be more adept at recognizing when everything is working smoothly, even the subtlest of changes. When friends get chronically ill and tell me they never knew anything was wrong, I always wonder, didn't they have even a slight indication that something didn't feel right? Wasn't there even a twinge? With an ability to scan your inner landscape, you gain the ability to intuit early when something is out of kilter.

If you learn how to look down that periscope, you'll know pretty much everything there is to know and what's coming around the corner. I always think it's like women who tell me they're surprised at a late stage in life to find out their sons or daughters are gay. I think, *Didn't you notice anything earlier? Were you asleep at the wheel?* I mean, there are clues.

Evolution Again

Neuroscientist Daniel Wolpert believes that the reason the human brain developed in the first place was to resolve the problems we encountered when we began to move around in our environment. Plants didn't have to move because everything was done for them: they're pollinated by bees, fed by the soil, rained and shined on. So why would they need to develop thinking? Animal behaviour hasn't changed much since the beginning; what worked for them then still works today. If killing, defending themselves and mating is all they ever had to do to survive, why change the habit of a lifetime?

We didn't get any of those animal add-ons they have – to climb, leap or stretch their necks to get bananas from the upper branches of the trees – so we had no choice but to learn to think. Now equipped with the new thinking expertise, we could build ships to import bananas from Jamaica. When we could no longer walk those challenging distances to find food, we invented cars to take us to restaurants. Now, with delivery services, we only need to use our brains to choose between the Peking duck and the sweet-and-sour pork with egg rolls on the side . . . oh, and seaweed.

Embodiment

By moving our focus back into the body, we can learn to bypass hurtful thoughts or emotions, nipping them in the bud before they do their worst. Usually, with mindfulness for emotions, you direct your focus to where you sense them in the body. With mindfulness for thoughts you pull focus to the body or the breath (a top-to-bottom approach).

Some people don't like to sit while practising mindfulness, but the great thing is you can do it while moving around in your daily life but still focusing on the body sensations. You'll become aware where those stuck emotions reside; maybe noticing that your shoulders are tense and held, your heart is pounding or that you are holding your breath. If you calm the body, it will calm the emotions, which in turn calm the thoughts (bottom-up approach).

Other people say they don't have time to practise

mindfulness and that, if they do have any spare time, they'd rather exercise their bodies than sit and do mindfulness exercises (which probably won't tighten their bum). Exercising aerobically does improve blood flow, strengthens muscles in the body and the heart and tightens the bum. We need to stretch our bodies so they don't seize up, and a flexible muscle is a happy muscle.

The Problem with Multitasking while Exercising

No question, all physical exercise is good for you, unless you go to extremes. (Many gym-goers have ended up in a neck/knee/back brace from over-pumping or overstretching.) Here's the deal: you don't have to stop exercising, just learn to do it mindfully by sending focus into the area you're stretching/contracting/moving. Not only will you be aware of any potential damage, you're also improving your brain. It's noticing that you're mind-wandering then bringing focus back to specific sensations in your body and doing it again and again that bulks up the insula and reduces stress. You *can* 'have it all': a tight bum, tum *and* a fit brain.

Things like t'ai chi and other martial arts incorporate and embody mindful movement. If you practise them, you become aware that your physical movements and your mental states are one and the same. But any activity – swimming, walking, lifting light weights, dancing – can be used as mindfulness in motion.

Nutrition: What Does It Mean?

I know, I know, I haven't mentioned diet. What you eat is who you are and how you are. This, unfortunately, is not my area of expertise, but there are about 2 million (contradictory) books out there that can tell you what to put in your mouth better than I.

I think, if it's green it's good; if it's chequered, don't eat it. I am in a continual state of confusion. I've gone through my vegetarian phase and, because I can't cook, I just ate nuts and berries, like a rodent. And I don't always trust food companies that shove the word 'organic' on the label. They're too expensive, it's like eating the Prada of the fruit and vegetable world. Fasting doesn't work for me, I get hungry. A few years ago, I just juiced. I shovelled pounds of vegetable and fruit into that blender and then vomited for a week. Someone told me later that, because the ingredients reduce so much, I didn't notice that I had drunk the equivalent of three football fields. Now, I'm on the Paleo diet so I eat everything that once had a pulse, hoof and horn. I've helped the world rid itself of cows because they're all going into me. So get out there and pick yourself up a book on nutrition and go nuts with the choice.

The Monk, the Neuroscientist and Me

It's time to turn to my experts, who may be more enlightened on the subject of the body.

Ruby: Ash, how do you see your body? How do you relate to it?

Neuroscientist: I grew up as a skinny Indian boy around a lot of big white Midwesterners. When I was young, I just wanted to look like one of the guys on the football team. I didn't like my body at all. I pumped iron and drank those disgusting protein shakes, and I even got one of those jackets with the high-school letter on the side.

Ruby: Those honcho jackets where they put the letter of your football team on it? That's sad.

Neuroscientist: Yeah, well, I did it so I could meet girls.

Ruby: Did it work?

Neuroscientist: Not really. I mean, the letter jacket was from the debate team, it didn't have the same effect as the football jacket.

Ruby: What about your body, Thubten? Do you like your body, if you can find it under those robes?

Monk: Well, my relationship with my body has changed a lot. When I was young, I was really into looking good, but I always felt sort of disassociated from my body. I was caught up in my head all the time. When I got seriously ill at twenty-one, I would talk about my body as if it wasn't mine. I would say, 'The heart isn't good,' as if it was someone else's. I think that's why I got sick. A total mind–body split. My appearance sort of fell apart when I became a monk. I'm quite chubby now.

Ruby: Do you miss being buff?

Monk: I think being healthy is good but being buff is pretty pointless. The mind is so much more interesting. Nowadays, my relationship with my body is about using it as a support for mindfulness, and that feels great. There's not much point having a 'hot' body and a rotten mind.

Ruby: Ash, I want to know how the body and mind are really connected. Is there some kind of spider's web made out of neurons from the head to the toes?

Neuroscientist: A spider's web is a great analogy, yes. When an insect gets trapped anywhere in the spider's web, every single string of the web vibrates. The spider can feel that vibration from any point on the web. The brain and the body are like that, they're totally interconnected, a single system. Anything that affects one part affects the whole system. It all vibrates together.

Ruby: So, if you stub your toe, how does your brain find out?

Neuroscientist: When you stub your toe, you activate pain-sensing neurons called A-delta fibres. Those send electrical signals up through the spinal cord and into the brain, but that in itself isn't pain. The brain responds to those signals with a network of areas, the somatosensory cortex, insular cortex and the anterior cingulate cortex, and those areas together generate feelings of shock, threat and suffering. It's those feelings that make up pain.

Monk: We talked about that before in the chapter on thoughts – the gap between perception and reaction, before you start judging the signals. This becomes interesting when working with pain. When you realize pain is an emotional reaction not just a physical one, you can do something about it.

Neuroscientist: That's a key point about pain. I totally agree.

Ruby: Okay, I understand physical pain but how do you deal with emotional pain? If you're scared, how does your body react?

Neuroscientist: Fear gives you an adrenaline surge – your muscles tense up, your digestion shuts down and you're ready to fight or run. If you unclench your fists, drop your shoulders and slow down your breathing, you'll feel a little less scared. The body and the emotion are one and the same.

In fact, Darwin used the word 'attitude' to describe how animals hold their bodies. An animal could be in a defensive posture, or an aggressive one, or an approachable one. Darwin thought that the posture itself was the emotion, not just an expression of the emotion. So it's a two-way street. Changing your emotions changes your body posture, and changing your body posture can change your emotions.

Ruby: So, if I'm hunching my back and baring my teeth, could I still like you?

Neuroscientist: Yeah, but when you fake a body posture like that, it won't feel natural. Hunching your back and baring your teeth makes you more likely to be aggressive, but it isn't a guarantee.

Monk: Our body never lies. Our mind can play all kinds of avoidance tricks on us, but the body will always tell us how we feel. It's important to listen to that. Sickness, for example, is a messenger – it can be a wake-up call to get us to see what's going on with our minds.

Ruby: What about with depression? I know when I'm depressed – I'm not skipping around in the petunias. I know my body is slow and my limbs feel like I'm lifting weights, but still I assume you can't lift depression by smiling.

Neuroscientist: That's right, it's not that simple. When you're feeling a bit down, not chronic depression but you're just in a grump, then opening up your chest, lifting your head and going for a walk can really help. But real depression is different. Body posture can be a start but it's not nearly enough.

Ruby: Okay, I get it. I always wanted to know, if you had a brain in a jar, could it feel emotion?

Neuroscientist: No, I don't think so. The brain needs the body to function, and the body needs the brain. It's a single system.

Ruby: But if I put your brain in a jar and sold it, would you be pissed off?

Neuroscientist: I would understand. I had the same idea for your brain. It's my retirement plan.

Ruby: Okay, now let's discuss exercise. I want to know why people are now beating themselves up for a six-pack, screaming like they're in childbirth for a couple of bulges. I mean, do you need a six-pack to sit behind a desk? I'd understand if your job was to lift the desk, but having a six-pack . . . does that improve your health?

Neuroscientist: Yes, there's no question that any exercise is good for cardiovascular health, and that's also going to be helpful for brain function. But people go to the gym and, while there, they watch TV or listen to their headphones. They go there to tune out so they don't have to think. That's very different from yoga, t'ai chi or most martial-arts practices, which put a lot of emphasis on mental focus. I think that developing an awareness of your body and how it moves is more important for health and even for strength than mindlessly pumping up your muscles.

Monk: Exactly. We could be totally distracted in the gym, running on that treadmill to run away from our minds, and also, when we're doing bench presses we might be unaware that we're dislocating a shoulder.

Ruby: Yeah, I've done squats and didn't realize I was giving myself a Caesarean.

Monk: Also, the idea of what's considered attractive changes with history and culture. Ages ago, and also nowadays in some parts of the world, being a bit fat was seen as a sign of happiness and success, while being skinny meant misery. Now, in the West, people are torturing themselves to achieve the ultimate 'thigh gap'.

Ruby: My mother always said, 'For beauty, you have to suffer.' That's why she put me in braces for forty-seven years.

Neuroscientist: But if the goal isn't just beauty but a healthy life and an active old age, bulking up without focusing on movement is a mistake.

Ruby: I know guys who did weights all their lives and, now they're older, they've literally turned into gorillas. They're all hunched, knuckles dragging on the ground. Who came up with the idea of the beefcake being the vision of health and virility? It doesn't arouse me. It's like a big slab of meat in trainers.

Neuroscientist: I don't know where that idea came from, or when exercise became more about vanity than about health. Most doctors would say that exercise builds muscle and improves bone density, and those are definitely good things. Actually, focusing on bulk instead of flexibility can be really unhealthy. Men may go for big chest muscles but, if they don't develop strong legs and core muscles to support the weight, they risk injury. Someone who has good flexibility and core strength may not look big, but they will be more fit. That leads to healthier ageing, fewer falls, more independence and less cognitive decline.

Ruby: What about people who can't stop pumping? They're obsessed with fitness and bore you to death talking about it.

Neuroscientist: Addiction to the gym is considered cool but, in the end, it's still an addiction. You're still chasing a high, looking for that right combination of dopamine and endorphins.

Ruby: We have wearables like Fitbits that give you a great high. It pushes you to constantly push yourself harder, like having a nagging mother on your wrist. One day, it's your best friend because you've done ten thousand steps and it congratulates you; the next, it's not even going to speak to you unless you give it twenty thousand. Eventually, it sues you if you don't climb Everest. You might be lying there passed out while it's still blinking at you to run up and down Machu Picchu.

Monk: Years ago, I remember reading an article about Demi Moore which said that she gets up at 4 a.m. and does sit-ups constantly until breakfast. Me and the other monks were horrified and said if she'd spend half that time meditating . . .

Ruby: She'd be enlightened by now. And while we're on Demi, where is she now?

Monk: I do think exercise is important, but there's so much focus on how we look, instead of what's inside.

Ruby: You don't win any gold medals for inner peace.

Monk: Sometimes, when I explain mindfulness, people who do a lot of fitness training tell me they can get the same thing when they're on the treadmill, so why would they need to meditate? I say to them, 'But you can't run all day. Or, when your boss is yelling at you, you can't drag a treadmill into the room.' If you're training your mind through mindfulness, you're learning how to lower your stress whatever the situation, you don't need to always be at the gym. Exercise doesn't train the mind. Yes, it will help you feel less stressed, but only while you're doing it, and for a while after, but not in the long term.

Ruby: But they can do both at the same time if they do mindful movement.

Monk: Exactly, there are methods where you use the body to train the mind.

Ruby: So you're getting enlightenment and a tight tush. Ha ha, Demi, who's laughing now?

You'll find the relevant mindfulness exercises for the body in Chapter 11.

5

Compassion

There's an expression, 'If you're shot with an arrow, you just pull it out.' It doesn't help to worry about who shot it, why they shot it and whether they'll shoot it again. Just take it out. (That's self-compassion, in my book.)

It's strange for me to be writing a chapter on . . . I have trouble saying the 'C' word, it's always been a hard one for me to say out loud because of my depression and the feelings of shame and self-disgust that come with it. It's probably because I stigmatize myself more than other people stigmatize me. I bring out the whip, thinking how self-indulgent I am, when other people have far bigger problems. I have to remind myself that these thoughts are symptoms of the disease. The biologist Lewis Wolpert writes, 'Thoughts are to depression what a tumour is to cancer.' But while I'm doing time in the darkness, those vicious thoughts seem so real and justified. This is why it's difficult to feel *self*-compassion, let alone compassion for anyone else.

My Story

I've been working on this book while on my *Frazzled* tour over the last year. I love to write on long train journeys, where I can finally focus (if I get bored, I can look out the window and see the odd cow), unlike

working in my home, which is the Piccadilly Circus of distraction.

At midnight on 7 February 2017, I was in a taxi coming back from Victoria Station, and writing in the back seat. Two hours later, at home, I realized that my computer was missing. Horrified, I remembered that, as I got out of the cab, my suitcase wasn't completely zipped up and I realized that my computer must have slid out.

A small note here: during a few of my more severe bouts of depression – and it could be coincidence – my computers seem to break down along with me. Either they suddenly go blank, never to return to life, or, on more than one occasion, I've spilt water on them and, again, kaput. In *A Mindfulness Guide for the Frazzled*, I wrote about a whopper of a depression I had a few years ago in America and its accompanying computer breakdown. No button on earth could bring it back. All that happened was, every once in a while, there were flashes of passing lights on the screen, like a UFO.

At the time of our breakdowns, (the computer and I) were touring the States, publicising my previous book, *Sane New World*. It wasn't going well, primarily because some publicity person who was supposed to be arranging my tour seemed to have a large number of screws missing. I ended up in the back of a vitamin store in a shopping mall two hours outside Los Angeles being interviewed by a bald person with three hairs glued to his forehead. His only question was, did I think green foods would cure cancer? (I don't think he had read the book.) That was one of the three interviews this PR person set up. (I don't think she had read my book either.) Another interview was with an eighty-year-old person who had never heard of me and asked if I knew a recipe for lamb chops. (She must have thought I was a chef.) There I was, in America, in a state of full-blown mental illness, wandering about with my lifeless computer. I spent my free time going from one Apple Genius bar to another, and not one 'genius' could bring it back to life.

So this time when I lost my computer in the taxi, it triggered the memory of depression and, once the monster awakens, he moves in. I called the black-cab lost-and-found helpline, which is the communications equivalent to a black hole. I went to the Apple Store Genius Bar in Westfield, begging them to please call iCloud and tell whoever's up there that I'd pay big money if they could just find my unsaved documents. They looked at me with pity. I was starting to go under.

The next day I got an email from someone telling me she had found my computer. She said she had bought it the day before at a street market. She must have opened it and seen the screensaver of me posing with the Dalai Lama. (Obviously, it put the fear of Buddha into her.) Anyway, I asked if I could meet her and copy my documents, telling her she could keep the computer. We met at an art gallery where she worked, about six minutes from my house. I offered her some money but she said all she wanted was to give me back my computer, telling me it would be bad karma if she took the cash. I couldn't believe it. I kept thinking, *What's the catch?* I'm a natural-born cynic; in my mind, no one does something for nothing. There's always a bill. She insisted on not just giving me the computer but giving me two works of art from the shop and asked whether I could say hello to her husband, the artist, on the phone. (I asked him if I could record our conversation.) He said, 'Maybe this was supposed to happen – you were meant to drop that computer and I was meant to give this message to you. You're honest and speak the truth. I adore you and what you're doing. Now you go and have the best day.'

My depression retreated into its cave and I revised my pessimistic view of human nature: there isn't always a price to pay when someone does something nice for you. So, with this mind-altering paradigm shift of an experience, I'll start my chapter on compassion. I no longer need to talk about the 'C' word, I can now say it or, in this case, type it: compassion. (See?)

Compassion: What is It?

So, what does it mean and how long has it been hanging around? Answer: a long time, or thereabouts. Around 1 million years ago, our ancestors were far more touchy-feely than we are today. Yes, they'd spear you if you stole their wife, but they also looked after their old and sick and the whole community looked after the kids. (Isn't that the dream . . . to have a village full of babysitters?)

Compassion comes from the Latin word *compati*, meaning 'to suffer with'. It doesn't just mean sending someone a Hallmark card with a baby pig wiping away a tear on the front and, inside, a message saying how sorry you are. That would go under 'pity' or 'patronizing'. With compassion, step one is feeling the pain of another and the big step two is being motivated to relieve it. It's the will to act that defines it rather than just 'feeling someone else's pain'. You feel you actually want to go out and do something about it. If I'm in pain and you just feel my pain, it's not going to help the situation. How can you possibly help me if you're in so much pain from my pain? Now I'll need to help you cope with my problem. Also, we sometimes jump at the chance to feel someone else's pain for the wrong reasons; we don't want to feel our own so we distract ourselves with theirs.

The benefit of practising mindfulness for compassion is that you'll be able to keep your mind stable in the fierce fires of someone else's pain without getting pulled in or overwhelmed. You'll be able to stand back, watching your thoughts and feelings, to make a clear, unbiased decision on how to help the other person by being aware when to say something and when to be quiet and just be nearby. Someone has to hold the boat steady when the storm comes in.

Empathy: What is It?

Empathy is a whole different ball game. It comes from the Greek word *empatheia*, meaning 'to feel into'. It's implanted in us from our mammalian forefathers, when we mimicked the facial expressions and gestures of each other, based on the monkey-see-monkey-do school of thought. When we mirror another's facial expressions, we feel what they feel, because our faces and our feelings are intrinsically linked.

Compassion is the feeling you get when you see someone suffer and this motivates a desire to help. Empathy is resonating with someone's pain but not getting it confused with your own. Empathy is unconditional, just as compassion is; we don't have to like the person we have empathy for, we just need to imagine ourselves in their shoes. You can take this even further: if you can feel empathy for someone who's harmed you, you're in the big league.

Self-compassion Comes First

However, if we don't learn to be compassionate to ourselves first, we can't feel it for anyone else. A mother has to teach her child to soothe themselves, but she can only do that if she can soothe herself; otherwise, there'll be two people drowning.

My opinion is that we project our thoughts about ourselves on to people around us (for example, I know that I'm a great liar and so I don't believe people are generally honest, and equally, if you have too much self-criticism in your head, you're spreading that abusive virus on to others). Conversely, if you're nice to yourself, you're probably generous and kind to everyone around you.

Somehow, we've got this idea of self-compassion confused with selfishness. It's far from being selfish because, if you can give yourself compassion, you won't drain other people by expecting them to make you feel good or, when you're beating yourself up, blame them for your bad feelings.

Learning to throw ourselves a bone of self-compassion increases our resilience and stability. When you have that security blanket of self-compassion, you feel like you can take more risks, come out of the box and be more creative.

Many of us judge ourselves by our accomplishments, ricocheting from feeling great when we succeed to sinking into misery when we fail. Our self-esteem goes up and down depending on what grade we've given ourselves. With self-compassion you learn that, if you do fail, it doesn't mean you're a complete failure as a human being, it means you screwed up on one thing. People who have more self-compassion find it easier to apologize and admit they've done something wrong when they've made a mistake. Their sense of self-respect isn't threatened because they don't intrinsically think of themselves as a bad person or a failure. If someone doesn't have self-compassion, they'll usually get furious if you point out an error because it stokes up their feelings of not being good enough.

Compassion to Others

A selfish reason to exercise compassion is that it makes you feel good. When you respond to your own or someone else's distress, you automatically go into caring mode, which promotes the release of opiates and oxytocin in our brain. Great friendships and relationships are the result of the exchange of these hormones, which create trust, rapport and closeness. The great thing about us humans is that we can learn to cook up those feelings.

If we get caught up in the habits of anger and fear, it's reflected in our neural wiring and we become trapped in that mind-set. In that negative state, it's impossible to pass or receive oxytocin. Our mirror neurons shut down and we can no longer interpret whether someone is trying to be helpful or critical, cruel or kind, leaving us feeling defensive, paranoid and unsafe.

Once we feel unsafe, there's no more Mr Nice Guy. We become terrified that, if we show any kindness, we'll be taken advantage of. That's why, in our culture, 'niceness' doesn't get high ratings. Toughness is in vogue and probably has been for a very long time. This could be why we have a fascination with other people's misfortunes, why the videos that get a billion hits on YouTube are usually of a baby falling into a chocolate cake or 'kitty gets flushed' (my favourite). In truth, we've always loved watching other people's pain, right from the eat-the-Christian gameshows at the Colosseum through to the humiliation and shame we see on *X Factor* which isn't a million miles from the sight of a lion eating a slave. At least the slave didn't have to sing. TV reality shows are based on slinging out the loser and cheering as they walk the walk of shame, out of the building, never to be heard of again. (Unless they humiliate themselves in some new and original way, maybe getting caught snorting coke off some politician's backside. That just might get them back in the limelight.) You can probably imagine that compassion doesn't win many viewers. This is why the news today gives you in-your-face close-ups of people buried in the rubble rather than a shot of the best damn apple pie in Indiana.

Neuroplasticity is Our Salvation

The neurons that wire together in your brain reflect how you sense, feel and think, second by second. They swap partners with every new input so that, with each new experience, your brain is rewired and gets redecorated, forming new neural patterns. That's neuroplasticity.

The landscape of the brain is constantly being reshaped. If you're watching a horror film, you can bet your booties the neuronal formations are reflecting it and your amygdala is up there pumping cortisol into every cell of your body, making you quiver in your seat. As far as your brain and body are concerned, fear is fear, no matter how you cut it: whether you imagine Freddy Krueger in your mind, see him on screen or he's actually in the room with you, the same brain states are activated. This happens with all your senses;

whether you actually smell something real or imagine it, it's all the same to the brain. Which begs the question, why bother cooking when you can just smell a picture of the food in a recipe book?

It must, logically, follow that when you experience or imagine something compassionate, the body and brain also reflect that state. The question is, do you want your brain to be a whirling dervish of madness or something that promotes ease, health and self-esteem? It's your call.

Compassion in the Brain

Rick Hanson, who I mentioned in Chapter Two, says that the brain can become 'hardwired for happiness'. Here are some suggestions to help you rewire:

- Help someone you don't know (unless they tell you to fuck off)
- Be happy for someone when they succeed (this one is a killer for me)
- Say sorry if you interrupt your husband/wife for the thousandth time (which I do on a regular basis – the interrupting, not so much the sorry bit)
- Let someone jump the queue in front of you (I know this is sacrilege in this country; just see if you can let them in without screaming, 'You asshole, who do you think you are?').

We need to learn how to 'do' compassion. It won't grow by itself in our neural jungle and, if we don't learn it, we'll go back to our very destructive and violent default mode (see Chapter One).

When being compassionate, there are no rules to follow. Any time you're moved to do something to help, that's enough.

Even if you don't do anything but be by someone's side and stay present in the midst of their suffering, that's enough.

My Story

I was in Cape Town and had been asked to teach mindfulness to young girls from the township who had been badly abused. As soon as I started, I could feel they were uneasy and that the last thing that they wanted to do was observe their thoughts. Mindfulness, in my opinion, isn't appropriate for severe trauma. When the trauma is resolved, or has eased off, you can try it, otherwise, I think you reopen the wound. I thought, *Drop it*, and for some reason instinctively asked if any of them had ever had a makeover. They never had, but the excitement ricocheted through the room. I came back the next day with my makeup. There they all were, lined up and totally focused, any sign of agitation gone. Here was something that made them feel important, as if they mattered. Simply by touching their faces to apply the makeup, their bodies relaxed, probably for the first time. When I did their lipstick, I touched their lips, which I'm sure would have normally flipped them out, but because I was gentle and had no other agenda, they softened and went silent. I almost cried, they were so still. When I finished, they all took selfies or posed together for photos (everyone has a phone in South Africa), like models displaying their beauty – and they *were* beautiful. I left feeling elated, and hugged them all, knowing this was probably the first time someone had touched them without taking advantage of their innocence. I loved those girls.

The Monk, the Neuroscientist and Me

Ruby: Thubten, you start with the definition of compassion.

Monk: Why am I the one who has to define it?

Ruby: Because you're the expert . . . It's your job.

Monk: I think the definition of compassion is that you're moved by the sufferings of others and you want to do something about it. So it's more than just a feeling, it's about action. Of course, it does start with deep feeling; it jolts your heart when you see someone else suffering.

Neuroscientist: That's right. Compassion incorporates the intention to act, even at a neurological level. Highly experienced meditators like Thubten activate their premotor cortex during compassion meditation – that's the part of the brain that prepares the body for movement. Compassion also activates a fronto-parietal circuit connecting attention in the parietal lobe to behavioural control in the prefrontal cortex and reward processing in the midbrain. That circuit is usually associated with rewards and positive feelings. Empathy, on the other hand, is associated with the insula and the cingulate cortex, and those regions are more typically associated with negative emotions. You're feeling the suffering of others and you're suffering with them.

Monk: Professor Tania Singer, the neuroscientist, did some very interesting research on this. She tested the brain activity of monks while they were asked to view images of people suffering, and the monks were asked to meditate on compassion. First of all, she asked them to only focus on empathy and, actually, their stress levels went up. Next, when they meditated on compassion, the stress went down and the areas of the brain connected with intention were activated. That dynamic mind state is sometimes even described in the Buddhist texts as 'bliss'. I don't mean they felt happy that others were in pain, I mean the intention to benefit others produced a lot of energy. This gives the flavour that compassion is something very strong and purposeful. It's not that you're looking at someone suffering and now you're suffering too, almost like you've caught an infection. That would be more like 'emotional contagion'.

Ruby: How would you go about training the brain for compassion if you're not a natural at it?

Monk: It's a step-by-step process. For most people, compassion is just a reaction to seeing someone suffering. The next step is to develop compassion as a state of mind that doesn't need an object to trigger it. If you train in this way, you'll want to help people in general, rather than just reacting to an individual case.

Ruby: Okay. At first, doesn't the training seem a little artificial?

Monk: When you first learn to ride a bicycle it feels unnatural but, if you persevere, it gets easier and you enjoy it. It's the same with learning compassion. Eventually, it becomes a natural part of who you are. Also, you begin to suffer less because you become less obsessed with your own problems, and that makes you happier. It puts your pain in perspective.

Ruby: Ash, if it's so crucial we have compassion, why aren't we born with it? Why do we have to train it?

Neuroscientist: We are born with it. As soon as babies are old enough to coordinate their arms, they reach out to pat and stroke people they think are upset. In the lab, you can put babies in front of a little stage show with puppets where the puppets either help or hinder each other. Afterwards, even three-month-olds will pick up the puppets that were helpful to others and they won't play with puppets that were mean.

Monk: So, does that imply we're hard-wired for love?

Neuroscientist: I'd say it just means that babies can make simple judgements about right and wrong, and that they have an instinct for compassion.

Ruby: I'm pretty sure I didn't do that as a baby. I would have liked the rebel puppet. I always think compassion sounds gooey.

Monk: Compassion isn't something soppy, it's actually very brave because you have to look at yourself honestly. You need to be willing to face your pain as well as the pain of others, instead of burying your head in the sand in avoidance.

Ruby: As a doctor, do you train at all for empathy or compassion? My doctor sees me as a piece of meat on a conveyor belt, especially the gynae. I feel like I'm on a YO! Sushi bar.

Neuroscientist: Yes, there's a lot of focus now on empathy training for doctors. It's not so straightforward, because you want your doctor to be empathetic but also dispassionate. No one wants their doctor to walk in the room and break down crying. Most patients want someone who cares very deeply about them but is calm and steady during a crisis.

Monk: Exactly, you're not just drowning in the empathy, you're helping them with compassion. The opposite would be the emotional contagion I mentioned before, which can eventually lead to 'compassion fatigue'.

Ruby: Thubten, what do you do to help people? I'm not trying to be provocative, I'm just curious how you would describe what you do as a professional monk.

Monk: I try to help people understand their minds through training them in mindfulness because, ultimately, the suffering we experience is related to our thought processes. So, if I can help people to work on their minds, I feel I've helped them find the causes of their suffering, and then they can start to transform.

It all begins with training the mind. I know it's an extreme example, but when I went on my four-year retreat, pretty much all the practices you're doing in there are compassion practices, even though you're not actually out there helping anybody, you're just in a room, but you're building that mind state that you can then take into the world. It's the same with everyone who even just sits down for ten minutes to practise mindfulness. During that time, they're not actually out there helping people, but they're building up the brain regions that will enable them to go out and take action.

Ruby: So, Thubten, what do you do when you go out there and act?

Monk: The work I do is usually about helping people with their minds, but I also help with charitable projects which feed those who are hungry or provide education and healthcare in Third World countries.

Ruby: I didn't realize that. I thought you just sat in a cave.

Monk: I also strongly believe that if you teach mindfulness to school children, businesspeople and politicians – those three areas – you have a chance to change the world's future.

Ruby: What about someone who's treated you badly? I'd like to be compassionate to my mother, but she didn't show me such a good time.

Monk: This is what compassion training is all about. That's when it counts the most.

Neuroscientist: So much of what stops us from being compassionate are our memories and grudges. The good news is that memories are re-formed every time we recall them, so if you've trained yourself to be able to access compassion, you can intentionally change the memories. The more compassionate thought eventually becomes the new memory. Of course, I know that as a scientist, but I wonder if I could do it with memories of my father, whom I never got along with?

Ruby: So, the story about my mother and his father completely changes. You're rewriting your history?

Monk: There's no point in creating a completely new story. You can't come up with a new story like 'I wasn't mugged', for example, when you were, but if you try and add a flavour of compassion, it feeds into the memory so you won't always have a trigger reaction of anger or revenge.

Neuroscientist: But should I just try to feel compassion in general? Do I need to forgive him, or can I just maintain a feeling of compassion?

Monk: What's the difference?

Neuroscientist: I suppose, neurologically, there is no difference. Maybe practising the feeling of compassion would infuse my memory of my father, even if I don't necessarily forgive him.

Ruby: So, if I look at the Disney film of *Bambi* when Bambi's mother dies (I can't even discuss it without filling a bucket with mucus and tears), if I picture that scene and then hold up a picture of my mother's face, does that mean I'll start to love my mother? Or start screaming when I see *Bambi*?

Monk: That's how compassion meditation works. You start by meditating on something easy that makes you feel compassion, to get the 'juices flowing'. For example, first you think about someone, maybe even a pet, whom you love, and then the compassion is activated. Then you imagine your friends and family and, eventually, you spread the compassion to the person you have a problem with, like your father, Ash, or, for you, Ruby, your mother. It's a step-by-step training process.

Ruby: So, if the juices are on, I can break the imprint in my memory?

Neuroscientist: I think you could. That makes sense neurologically.

Ruby: So, you're tricking your brain, or outsmarting it, in a way?

Neuroscientist: I don't think it's a trick, any more than learning to throw a baseball better by changing your grip is a trick. It's just learning how the body works. Emotion and memory are two sides of the same coin. When you're recalling an event, the limbic system replays the emotions that go with that memory. If you change the emotional context, even if it is a trick, then the memories are going to be recoded. You are not changing the memory itself, but you're changing your emotional response to it.

Ruby: How many times do you have to do it?

Monk: A lot. You couldn't do it just once.

Neuroscientist: I think of how many times I've practised it the other way! Remembering and going over my negative story with negative emotions.

Ruby: So, you're saying, if Ash practises compassion, someday he'll be able to look at a photo of his father and feel the old *Bambi* juice automatically?

Monk: That's the idea.

Ruby: So, let's say I see my mother with compassion evoked from the *Bambi* story but then I start remembering things like she made me clean the shag pile carpet with my tongue, it's not going to match up with the feeling of compassion when I picture what really happened or the stories I've been telling about her. Do I have to sieve through every story and add compassion? It will take years.

Monk: You don't have to reprogramme each story, you just have to reprogramme your story of who she is, so the memories will also change their texture.

Neuroscientist: Right, that's exactly it. If I can stop my automatic response to my dad, then I start to feel how hard it was for him. I can replace my reaction with some understanding of how he might have felt, so the memories are a little bit different. It's the forgiving that changes your brain.

Monk: And that would be compassion.

You'll find the relevant mindfulness exercises for compassion in Chapter 11.

6
Relationships

I'll be honest: I don't really know what constitutes a successful relationship. We don't entirely understand the workings of why we choose whom we choose but, happily, it's not just me who's in the dark. None of us knows. In Chapter One, I mentioned that much of what influences our decisions is not rational thinking but the ancient whispers from our past. It is probably a waste of time when you're describing what you're after on a dating site and you write, 'I'm looking for a sunny, fun-loving Virgo who likes laughter and jumping in foam.' You probably aren't aware that, underneath this conscious thought, you might actually want a hunk of man-meat who'll drag you home by the hair. Another possibility is you're asking for something that does not exist on this earth, like a 'sensitive, good-looking, billionaire'. No matter who you think you're looking for, it's your biology that ultimately makes the choice.

In the old olden days (which my daughter thinks is when I was young) – I mean, thousands of years ago (this is when she thinks I was young) – most people chose a partner because they made them feel safe, or in order to get the right bloodline (royal) for keeping up with the A-listers on the family tree. In our world, we don't think it's enough to marry someone for companionship, we need to find a Mr or Mrs Right. More than a few men I know out there want a partner who can do what no woman has ever done successfully: take care of the kids, be sociable, successful, sexy, savvy, skinny and know how to cook. (Ha ha, I say. Good luck.)

It's All in Our Biology

The job of an evolutionary psychologist is to dig into the primitive biases that affect our judgements, decisions and choices. I met with a very accomplished one called Andrew Dellis, and he told me about the results of research done on identifying our more primitive drives. (I found it deeply disturbing.)

He said we need to understand that we're motivated by biological forces, and this will then allow us to think twice before reacting blindly to a situation. A little knowledge of human behaviour means you might make a better decision, and also give yourself a break if you choose a partner who's a dud.

In one experiment it was found that women are more attracted to the alpha male when ovulating because of an innate desire for strong genes. This means they'll be more attracted to guys with masculine, symmetrical faces, because this, they say, is an indication of healthy genes.

It appears that this type of guy hooks the innate desire for what's called 'maximizing inclusive fitness', which refers to maximizing the number of offspring likely to survive, ensuring our genes are passed on. (Strangely, I don't think of that when I see those gorilla men.) This alpha comes with a full tank of testosterone, which is supposedly an indication of a strong immune system, meaning a greater chance for the kids to inherit it. Sadly, this *über* male also comes with a tendency to be dominant and aggressive. Maybe so many women stand for that macho bullshit simply to get some of those *über* genes.

Evolutionists suggest that when a woman isn't ovulating she'll be attracted to a more caring, feminized type of man, maybe less buff but more secure. One of the reasons she might consider the safer male is because she has a limited amount of eggs and the gestation is relatively long so she might want to find a partner who'll stick around until the offspring leave home. (These days, that can be up to about forty-five years.) Men don't need to invest in the one fertilized egg because they can spray thousands more, right up to the grave. If the human baby did what other animals do – swim or fly away after birth – both males and females could go off and find someone else if they chose and life would finally be fair.

All I'm trying to say is, if you're going to get married to someone, make sure you don't meet him when you're ovulating. You might just have been blinded by lust and he will not be a stayer.

Another experiment with a group of women involved giving them men's T-shirts to smell. On average, they were more drawn to the smell of the men who had a stronger immune system. Apparently, you can smell that. Imagine the sales figures if you could buy it in the shops? (Eau de T-shirt.) The lesson here is think first, smell later.

Men don't know what they're looking for; the explosion of images of fantasy women (see porn and models) bears no relation to actual women. This could be why men feel disappointed when they find out you have imperfections, like body hair.

My friends always want to know, given the choice, do they go with the roller-coaster, passionate sex addict or the guy you can watch TV with? (Again, my advice is not to make your choice of male while you're ovulating – you may make a terrible mistake.)

There are species that have the answers. Take the bonobo, the most liberal animal of them all. The females have sex with males only when they're ready to replicate and roll. The males engage in something called 'penis fencing' to win over the girl or, if they're gay, just to enjoy it (like penis fighting is going to turn the girls on?). When the girls aren't breeding, they spend their time having sex with everyone, willy-nilly, male or female . . . all the time. What's wrong with them as a role model? I'm just saying, there are other ways to skin a cat, sexually.

My Story

Perhaps Ed just had the right smell at the right time. His smell seems to have lingered, because I haven't gone on to sniff anyone else. I thought I had chosen Ed for more subtle reasons, i.e. his length (long legs), so that my children and their children would get more than the stubs I inherited, which eliminated my career as a showgirl and catwalk model. This feature Ed has is also what's known as an 'evolutionary advantage', in that the *Homo sapiens* with the longer legs could stride over larger areas of terrain to find food. We short, stubby ones fell behind but, luckily, my people were funny, so probably someone with legs gave us a lift on their back (that's my interpretation of why the Waxes exist today).

Also as far as I'm aware, I chose Ed because he had Grade-A sanity genes and I felt that would break the chain of thousands of years of Wax madness in one fell swoop. (I chose correctly, because none of my three children take crack, which I assume defines them as normal.) We've been together twenty-eight

long years. Many people would say that it's a miracle. I mean, if Brad and Angelina couldn't keep it up, no one can. You'd think adopting a child from every Third World country, all of whom they no doubt love, would be enough glue to keep them together. Nope. And it didn't work out so well with Mia Farrow and Woody Allen either (I made sure we never adopted a good-looking Vietnamese child who could iron clothes, and that Ed might run away with). It all depends on what you mean by a successful relationship, and I'll bet no one really knows. Love comes in many packages and we experience it in different ways; some need the fire-works of high hormones, others just want someone to face them while they eat so they don't feel lonely. There are people you hear about who are crazy about each other to the very end, dying together on a park bench while holding hands. I have never met them but have heard through urban myths that they exist. I'm sure if I interviewed them I'd probably get bored, so I'm sticking with those who are still floundering.

What Goes Wrong in a Relationship?

According to evolutionists, although we're a pair-bonding species, we're not built for the long haul. We're only monog-amous, from a biological standpoint, up to when our children are able to survive on their own. These days, kids don't leave home until they're forty-five, as I said, so now some of us stay with our spouse a horrendously long time. There's a comedian who once said, 'My wife and I had thirty wonder-ful years and then we met.'

I heard someone else say that, unless you work on a rela-tionship, it will die (unless you have no disagreements . . . and if you claim that, you're lying). Obviously, any relationship changes as it goes on but, if you stay curious about your partner and keep the communication lines open, it will last. If you find you have nothing in common, in my opinion, there's no shame in calling it a day. Some unhappy couples really believe they have to swallow crap till death do them part. (I'm talking about open warfare, not the normal bitching that comes with the deal.) If you have kids, it is more complicated but, even then, you have to weigh it up and decide. Will the damage be worse if you split up or if you stay?

There are also relationships where one or both go quiet, stonewalling the other, thinking if they keep schtum, they'll have a quiet life. This tactic never works because some day one of you will detonate and the explosion will be heard for miles around, especially by the kids. Don't think they're not picking up every last scrap of hostility. You can't kid a kid.

I also know people who married someone they loved but, many years later, they say that their partner is no longer the same person they once knew. Why is this such a surprise to everyone? Every cell in your body is completely replaced every ten years by new ones – what are the chances the two of you will still have things to talk about? Your new pile of cells may not like the other person's new pile.

With relationships, we sometimes choose an innocent bystander to pin 'the perfect partner' label on and then, a few years afterwards, we blame them for not being who we presumed they were. Mostly, we aren't aware why we're attracted to someone and it could simply be because they remind us of another person. I have a friend who married someone who had Tom Cruise's nose. Years later, things fell apart, because he didn't really act like Tom Cruise (as seen in films). You can't divorce someone because they didn't live up to their nose.

Whoever you happen to end up with, they cannot fulfil all your needs, because we are not one single entity, we have many sides to us. In a single hour, I can be childish, tyrannical, deep, shallow, deceitful, manipulative, compassionate . . . the list is long. I change depending on who I'm with, where I am, the time of day, my hormone levels and how hungry I am. We are all in a constant state of flux, so what are the chances of being in sync with another person for longer than a few minutes? It's a miracle that marriages last longer than half an hour. In certain moods, Ed is a joy to be around; in others, I could happily club him, especially when he eats.

One thing that makes our relationship work is that neither of us thinks of ourselves as a couple. We are two individuals who happen to laugh at the same jokes and have each other's back. We don't think of ourselves as an 'us'; we are each an 'I'. And we don't harp on about how different we are from when we first met. It's a hard, cold but true fact that everything and everyone changes. We have to acknowledge the fact that the only thing that's constant about anything is change. Our bodies and minds are more verb than noun; they flourish, erode, age, replenish and gain energy.

Ed and I know we're different people now. Time moves on. You can't blame each other for that.

What to Expect from a Partner through the Ages

One thing that helps when you're choosing a partner is to be aware of where you are in your life and to imagine how that might change. There are certain set points in life where your interiors and exteriors have a rehaul. Many of us don't want to know this, because we don't like to notice change but, just like puberty, which is undeniable, we have other extreme biological tectonic shifts throughout our lives. I'm here to tell you what they are before they take you by surprise. It may sound cold but all of this is to help you and I've found no one ever tells you about these changes, so I'm going to. I wish someone had told me.

18–25 years old

For both men and women: follow your hormones during these years. Have as much sex as you can (not forgetting birth control) but don't make any commitments because, right now, your biology is in the sex driver's seat while your mind is out of town. Don't try and look for your underpants the next day, you will never find them. Keep in the back of your mind that the drip feed of passion will, someday, sooner rather than later, dry up. Don't confuse a shag with being in love. It's confusing because oxytocin is released in an orgasm but is also associated with love. So if he/she doesn't call you again, just say to yourself, 'I just feel this way because of a release of oxytocin so, no biggie.' Basically, ignore your mind and let your genitals be your guide.

25–30 years old

For women: you have five years where you can, guilt free, build a career or travel. After thirty, if you have children, you'll always feel guilty, whether you choose to stay at home with the baby, giving up your career, or go to work and leave the baby. It's a lose–lose situation. Take advantage of these five years: you will never be this free again.

For men and women

You should still have as much sex as possible because the hormones are up there bubbling away. Also, this is probably the time you'll have your best body, so throw it around as much as you can before the big flaws start and gravity does its worst. I don't want to discuss what those are; you'll find out.

30–45 years old

It's almost impossible at this point to picture that, someday, something called menopause will fly into your window (for the men, it's penile dysfunction). If you're in your early thirties, forget I said these things and just have a great time dancing, laughing, having sex in lit rooms, drinking all night and not getting a hangover . . . I am getting so depressed writing this, I'll move on.

Around 34

For women: if you're deciding on a partner, be aware of what kind of life you want. Do you want the successful alpha who may be able to accessorize you until you're about fifty-five and then it's probably bye-bye, because alphas are in great demand and there will usually be younger female spares waiting around to overthrow you? Look out your window: they're circling like vultures at this very moment. Some of them are just being born.

For women

Alternatively, if you want to ensure that, when you get older, there will be someone around to clean out your bedpan, go for the more feminized, nice guy. He may not be a stud but he won't mind changing your nappy. You may get bored, but who else is going to do this job?

For men

You don't have to make any decisions. Just go on working and playing. If you have any guilt about the kids, your wife will feel it for you.

By 45

For women: if you were thinking of having kids, think again – well, maybe don't think because, chances are, you're low on eggs. Now, you should start to think about what a great life is waiting for you without children. You'll be free to do what you want, when you want and not have to think about someone else all the time. The mistake people make at this age is to get a cat or dog; then they're slaves to their pets and can't go out for fear that the pet will have a heart attack in their absence. Single life is a great life, but it's better to decide before forty-five so you're not taken by surprise. I have many friends who, like myself, refused to acknowledge that we age . . . but, in some areas, it's not up to us; our eggs, or lack of them, will decide.

For men

Just keep going the way you've been going. You can get married at ninety and still breed, so there's no rush.

45–50 years old

For men and women: if you're not married at this stage and want to be, I suggest going on the lastminute.com dating site, or just find anyone with a pulse and throw a ring on it. The good news is, a culling might have happened where one of your married friend's partners has died, so now is the time to swoop in there and get the living one. It's a second-chance time for old newly-weds.

If you're married, you've probably had a free ride to this point because, what with the kids, you don't have to pay so much attention to each other. So far, you've always had something to talk about: the kids. You'll complain about the 'terrible teens' and roll your eyes, wishing for the day they grow up. Be warned: when they finally do grow up and leave, you'll look at your partner and probably have nothing to say. Kids are the world's greatest distractions while you've got them but always remember that, someday, they will go. I know this is almost impossible to imagine, because we don't know what's going to happen until it hits us in the face. I've found that planning is power but their departure will still kill you.

My Story

At a late age, while my kids were taking their A levels, I enrolled at university. This would guarantee that I would be the one leaving with my suitcases packed and they would be left sobbing in the doorway. When I'd come home to visit, they were always happy to see me, and even learned to cook for themselves in my absence. Two years passed, and I passed. It's all worked out; all is forgiven now. They even came to my graduation and filmed me getting my Master's at Oxford. They looked so proud watching me from the balcony of the Athenaeum as important people spoke in Latin and I had to bow every once in a while. (It was one of the happiest days of my life.) My kids said everyone else looked sombre, but I was smiling and beaming so insanely that I looked like a demented clown head. I posed with my family in my bat gown and square, tasselled hat, like you're supposed to when your kids graduate. Basically, when they left home for university, I was already gone. It was win-win. I'm smart; they're smart.

For women

Once the kids are gone, you may find, if you've stayed at home and given up your career, that you've lost your mind (not all of you but, in my experience, quite a few). No one warns you about the empty-nest thing, and then it's there and you better have a safety net. This is the stage where the chance of divorce is at its peak; around 35 per cent of couples throw in the towel. At this point, the wife has nothing to talk about except her new hobby, making pots out of sawdust, or the husband wants out because he wants to be young again. The wife can't make that happen, but a twenty-five-year-old can. Don't worry, later, she'll get dumped too, because he'll keep getting old so he'll have to keep recycling the women for a newer model.

50–65 years old

For men and women: if you're married and have been for a long time, you'll notice that the stories are on a loop; you've heard them all before. You will deliver the punchline of jokes before he/she does. Inevitably, you've both run out of material. For women: at this point, you can stop shaving your legs and start gaining weight without fear – unless you're with an alpha, and then I'd say keep shaving and throw in a facelift. If you married an alpha, at this stage, he's either long gone or going tomorrow, so don't make plans for dinner. If you're with a nice guy, he won't notice or mind you turning into a fat carpet.

65–99 years old

I have no idea. I assume, if you're with the nice guy, you don't even have to sleep in the same room, eat in the same room or even breathe in the same room. When you get seriously old, you won't even know he's there. If you are still in love by this age, still exchanging large amounts of oxytocin, you've won the jackpot. This makes the Oscars and the Booker Prize pale into insignificance. Very few enter this sacred realm of being in love and faithful as an octogenarian, but we can all do it if we learn that life is a compromise and we are containers full of contradictions. We want everything – danger and safety, company and solitude, talking and silence, freedom and boundaries – and if we can resolve that in our heads, we won't blame the other person for making us feel short-changed. You are the interior decorator of your life. (I made that up.) Happiness is wanting what you have, not having what you want. (I didn't make that up.)

The Monk, the Neuroscientist and Me

Ruby: What do you think of the expression I just made up? 'You are the interior decorator of your life?' Would you buy the T-shirt?

Neuroscientist: I think it's great, Ruby.

Ruby: No, you don't, you're lying. I know you. Okay, moving on, I'm sure from the neuroscience point of view, in terms of relationships, it's our biology that drives us.

Neuroscientist: 'Relationship' is a word that scares scientists, so let's call it mating behaviour. We can look at three systems in the brain that drive us. I'm simplifying it here, but it's helpful to break it down in this way. The first system is lust, and it's driven by adrenaline. This is a purely sexual drive; it's intense and short-acting and then it wears off.

Ruby: I'm sure with lust it's like any grade-A drug, you have to keep upping the dose to get the kick. Eventually, you'll need to bring in the handcuffs and items of a crotchless nature to keep the flag flying.

Neuroscientist: I wouldn't know.

Ruby: Yes, you would.

Monk: I wouldn't know either.

Ruby: No, you wouldn't.

Neuroscientist: The second system is involved in romantic love, and the main chemical at play here is dopamine. Dopamine is part of the brain's reward system, so it's a strong driver of human behaviour. That's why romantic love is so compelling. It's like a teenager's view of love; it's about obsession.

Ruby: It's more about poems than putting out.

Neuroscientist: Exactly. The third system is attachment, or pair bonding, and its main component is oxytocin. This kind of love isn't obsessive; you can think about other things besides your partner. It's motivated by caring for the other person, rather than devouring them. These three systems should work in balance with each other, on different timescales. Adrenaline works for minutes, dopamine lasts weeks to months and oxytocin lasts from months to years. Oxytocin causes changes to brain connectivity over time, but even that fades eventually.

Ruby: So, what do you do when everything fades? A friend of mine wants to know.

Neuroscientist: Well, you can boost your oxytocin levels with physical touch and orgasm.

Ruby: But, eventually, you'd have to stop because you'd get vagina fatigue.

Neuroscientist: I hadn't considered that.

Ruby: What if you don't have many orgasms? Do you still stay bonded? Friends of mine want to know.

Monk: I think oxytocin can only be maintained when love is unconditional and less bound up with self-interest. But how many people truly experience that? There's so often an undercurrent of an agenda, or a need for validation from the other person. So, an important point is to move beyond ego within the relationship.

Neuroscientist: That's exactly it. If couples can build a relationship, a real companionship, they can sustain higher levels of oxytocin over time.

Ruby: Ash, what's your ideal relationship?

Neuroscientist: Well, as a scientist, I'd say a couple of weeks of adrenaline then a couple months of dopamine and, eventually, a lifetime of oxytocin.

Ruby: Thubten, are you just an oxytocin guy?

Monk: Yes, I think it's much more sustainable and meaningful. It's more to do with love than attachment.

Ruby: What do you think is the difference between love and attachment? Chemically, they seem to be similar.

Monk: Attachment is often based on need. The problem is the more we grasp, the more deficient we feel. It's interesting how people say, 'You complete me.' Aren't you complete? If you need someone to complete you, then you're making yourself less than you are and you'll always feel there's something missing. But if people can share happiness, that's a different story. Then it can work. Actually, there's no such thing as a perfect relationship. All we can do is make the decision to put the work in. A relationship isn't a 'thing', it's something you 'do'.

Neuroscientist: This is one of those places where an understanding of brain function can help people. If you know about the neurochemistry of love, you'll be less surprised and disappointed when the adrenaline and dopamine start to wear off. You won't think something tragic has happened and the relationship is doomed. It helps you to accept the evolution of love within a relationship, to know that it's going to change and that change has a chemistry to it; your biology ebbs and flows.

If you ask people what their ideal of love is they'll usually describe a kind of Romeo and Juliet relationship. So many books and films reinforce that: we're told we should be swept off our feet and that Celine Dion should be running through our heads on a twenty-four-hour loop. So, when that phase ends, people think the relationship is over. Maybe they get divorced because they can't recapture that feeling. But the head-over-heels feeling has nothing to do with the other person, you just want another hit of dopamine. It's an addiction.

Ruby: Thubten, you know you said a relationship is about work? A friend of mine is asking, what's the work?

Monk: I think the work is about becoming more self-aware, blaming the other person less and having more compassion. It's about mutual respect. If we don't start owning our stuff, the whole thing starts to crumble when the chemicals die off.

Ruby: It's so hard to respect someone when everything that person does pisses you off. And even if he's not doing something right now, you go into your bank of memories and remember what pissed you off in the past. Ed can say to me, 'Stop using that tone of voice.' Sometimes, I'm actually not using that voice . . . though most of the time I am. Ash, did you have that Romeo and Juliet thing when you met your wife?

Neuroscientist: I was never a believer in that thing before, but yes, I had it when I first met my wife. I walked into the room and immediately wanted to be with her.

Ruby: So did you think, as a neuroscientist, *Oh, I know what this is, it's my adrenaline and dopamine?* Did you analyse what was happening to you?

Neuroscientist: No, I didn't think anything. I just did everything I could to flirt with her.

Ruby: I would think, as a neuroscientist, you'd want to know why you feel what you feel?

Neuroscientist: I'm definitely curious but, when life is happening, neuroscientists have dumb human reactions like everyone else.

Ruby: So, you're unaware, like the rest of us? Do you know what's attracted you to other women in the past?

Neuroscientist: With one ex-girlfriend, it was her ass.

Ruby: I thought you looked at the brain, for God's sake. That's your specialty, not the other end. What was it about the ass?

Neuroscientist: I don't know, I guess it was shapely.

Ruby: You are so shallow . . . was it pert or what?

Neuroscientist: Just shapely! With my wife, Susan, it was also a chemical thing; either she drugged me, or it was some mating chemical that made me want to have a child with her.

Ruby: Why do you think we pick one partner over another?

Neuroscientist: There are probably lots of reasons we choose a particular partner, but that initial instinctive draw might have to do with evolution and genetics.

Ruby: So, we're not just attracted by the smell of an alpha, like I was told?

Neuroscientist: No, it's not all about alphas and betas. Genetic studies show that people tend to be attracted to partners who complement them on a genetic level, who have a set of genes that they lack. It's magazines and television that teach us that only one look is attractive, even though biology pulls us towards diversity. I know a tall, beautiful woman who has always been attracted to short, bald guys. Her attraction to that type of guy is automatic and, to me, that's biology at work. But you don't have to be opposites, it's just that people tend to be attracted to mates with genetic traits they are lacking. That makes for healthier children with stronger immune systems, but it won't necessarily produce a good marriage.

Ruby: What if you don't want to have children, or can't, or you're gay? Are you still choosing a partner based on biology?

Neuroscientist: Yes, surprisingly, there is evidence that gay people are also attracted to genetically diverse partners. It's about biology driving attraction, not just reproduction. Your genes are trying to pair you with a mate, but that doesn't mean you're going to have children.

Ruby: Can I go back to this finding someone who's genetically different from you? This couldn't be more true with Ed and me, as I have never even seen Ed's face; he's six feet two and I am a toadstool. Completely not in my gene pool.

Neuroscientist: But *if* you could see Ed's face, I think you'd get along. It's clearly worked out, because all your kids have the right numbers of fingers and toes.

Ruby: Thubten, how have your relationships changed as a monk?

Monk: Before I was a monk I think there was quite a mercenary quality to my relationships. Looks were the currency and it often felt like a situation of buying and selling. I did have some long-term, more serious relationships, but I would get very attached. It was all so intense and I would become quite obsessed.

Ruby: Did you ever stalk anyone?

Monk: Yes, I've been known to stalk if a phone call wasn't returned. Now, as a celibate monk, and with my mind in such a better place, I find that I can have so many warm and loving friendships and, because there's no agenda, I have relationships that feel so much more spiritual and enriching.

I'm not saying celibacy is for everyone – for one thing, the human race would die out! But it's very good if you're quite an intense person like me. I became a monk, where you vow to be celibate, so I'd have less distractions, which means I can wholeheartedly serve others.

You'll find the relevant mindfulness exercises for relationships in Chapter 11.

7

Sex

The Monk, the Neuroscientist and Me

Ruby: Thubten, when was the last time you had sex?

Monk: Twenty-five years ago.

Sex

Ruby: Well, that about covers it.

8

Kids

Babies: In the Beginning

I wrote about babies and how to grow them successfully in my last book, *A Mindfulness Guide for the Frazzled*. I can offer no better advice in this book on the topic so, if you want to know anything, go and buy my last book and turn to the chapter on babies.

Just a refresher: When your baby comes out of the float tank known as you, it's sprouting 25,000 neurons per minute and, as they branch out, they make 2 million connections per second. Now, don't be frightened by what I'm about to say, but every interaction you have is connecting those neurons, building the mainframe of your baby's brain. You are the sculptor and your baby is the mound of clay. Luckily, as soon as they're born, you'll get a full complimentary set of hormones to help you know what to do. Out of nowhere, you'll suddenly speak a new language called 'Motherese' and words like 'Woojie' will spring from your lips. Don't be alarmed: it's not you losing the plot, it's the hormones talking. From this point on, you and your baby will become mirrors of each other, reflecting each other's facial expressions, movements and, later, emotions and words. Our facial muscles are directly wired to our brains via nerves, so every expression sets off different cascades of hormones, evoking specific emotions. If you smile, your baby smiles and feels good. If you continually make a mad face, guess what happens? Your baby will someday need a shrink . . . for many years – and guess who's paying?

You will become a single unit, symbiotically responding to each other, each of you trapped in a reciprocal tango that never ends. And this dance is what ultimately lays down your child's characteristics. They inherit specific genes but everything that happens to them, starting with your interactions, will strike or silence those genes, determining their future abilities and traits.

Undo the Damage

And if they do get into bad habits, the good news is that they can be unstitched later in life, thanks to neuroplasticity. If you change how you see and feel about the world, your genes change (it's called epigenetics, for the fancier of you readers). It's never too late to rethink your thinking. This means, even if your parents screwed you up, nurturing experiences later in life can dramatically revise your biological makeup, triggering different genes to switch on or off. It's not too late to change your and your baby's gene expressions. Simply read this book and follow the instructions in the mindfulness chapter and you can repair the damage.

A psychoanalyst named Donald Winnicott came up with the expression 'good-enough' parenting. Thank God I'm off the hook; I had very little maternal juice. For example, I didn't know that excessive crying meant the nappy was filled to capacity. I thought Max was just in a bad mood so did some clown faces at him. He eventually exploded. (I'm not a mind reader.) Winnicott said that parents need to fail sometimes, to make mistakes when they try to fulfil their baby's needs, so that the baby can experience frustration. It teaches them how to 'man up' later in life and deal with frustrations successfully. If the parents try to be perfect, coddling them and never saying, 'No,' it can screw them up; they'll assume someone will always be there and give them what they want. So, it's important for parents to make mistakes. Yay!!!

Remember: Those who are nurtured the best survive the best.

Parents and Kids

You know your baby has become a child (around age four) when they start asking questions and demanding answers. You better have some ready. One of my kids asked at that age, 'Mommy has a front bottom' (female genitalia), 'my sister has a front bottom, so why does daddy have a chicken-leg bottom?' I wasn't ready for that one and, in those days, you couldn't google answers. By four, children are developing more solid personalities and you can no longer treat them as an extension of you. This is a delicate time because some parents (like mine) start getting into conflicts because they assume their kids are just bigger versions of the babies they once were. No, kids by this age are now individual people and you may find you have nothing in common with them. You may even dislike them, but it's too late to send them back.

Know Your Luggage from Theirs (Things I Wish I'd Known Before)

If you don't become aware of your habits of thinking, feeling and behaving, you'll pass your crap on to your child. How do you get that awareness? I hear you ask.

1. When you feel as if you've been stabbed in the heart because your kid has hit your emotional bullseye, they probably didn't do it on purpose, though it feels that way. Most kids don't start off spiteful, it's something they learn later in life. When you get the knife in the heart, register it and, even if you have to leave the room, don't respond; wait for the hot reaction to pass. If you can notice those moments when your trigger's been hit and manage not to snap at your child, that is emotional intelligence at its finest.

   When your mental storm has calmed down, talk to your child about what happened without blame and without sharing your pain. Maybe, if they're old enough, tell them that what they said or did sparked off a memory for you. 'Mommy is very fucked up,' is something I say a lot. This helps them see that you're human, which will surprise them.

2. Another way you might inadvertently pass on your luggage is by passing on your fear of failure, especially when they start going to school and getting marked on how smart or stupid they are. People I know who were relatively cool about life suddenly, when they have kids at school, will lie, cheat and stab anyone who gets in their way of getting their kids into a better school. They lose perspective and don't realize that when the kid is forty, no one will ask what school they went to. Except people who go to Eton (which should be barked like a peacock when said: 'EEEEEEEE-TON!!!!'). They will mention they went to school there even on their deathbed.

My Story

One of my daughters is an actress. Each time she goes for an audition, I try, I try . . . but each time she comes back, a pathetic, feeble voice squeaks out of me: 'Did they like you? Did they laugh? When will they tell you? Were you good? Did you feel like you did enough? Can I call someone to find out how you did?' This is said as one long, desperate sentence. My daughter, meanwhile, had a good time and thinks if she's right for a part, she's right; if not, so be it. I am one hundred years older than her and have not got this into my head. I used to take it all as a personal rejection, hungrily seeking the approval I never got at home. Every time she auditions, I relive the horrendous memory of when I auditioned for the Royal Academy of Dramatic Arts. I'd just finished my audition, which was met with stunned silence, so was standing with other hopefuls, waiting for the names of the 'call backs' to be read out. I had already been to six prior auditions at the Royal Academy, and my name was never read out. I was now at my seventh audition and, again, my name wasn't called. I asked in a desperate voice, 'Are you sure you don't have my name on the list?' And the man, who I will never forget, said in that dead, English voice, 'Yes.' I kid you not, I went back to my flat and took six Valium to kill myself. (I only had six.)

Each time my daughter auditions, I still feel the spear of agony pierce my heart from the memory of my name not being called. Now, because of mindfulness, I've learned to (sometimes) hold my tongue and even remove myself from the premises when she

comes home, because I still can't stop my face looking puckered and desperate. I can't get it into my head that, with all that rejection, I still did all right – more than all right. But we mostly remember the negative things, because the lance is much sharper when it's cruel than when it's a nice lance. I know my lances intimately.

3. Be aware that, from babyhood onwards, when you say something, your kid will not just hear your words but pick up your feelings underneath. If you happen to wince whenever you're talking about spinach, spiders or your mother-in-law, they'll also start to wince around those things. If you're not aware, your feelings about practically everything will be encoded in their brains, and they won't know why they've ended up scared of spiders, hating spinach or thinking Grandma is a bitch. And you don't just pass on the wincing, you pass on your feelings of inadequacy, distress, dread, rage, shame, and so on. If you've passed on some of these attributes, remember it's never too late to rewind and recode for both of you. Once you're aware of what you're passing on, everything changes. Insight is the name of the game.

4. For all you parents who are nervous about how smart your kids are, my tip is, rather than forcing them to learn by whispering Mandarin into their ears while they sleep, focus your attention on building your relationship with them so that, when your child feels hurt or stupid, they'll come to you, knowing you can handle it without flying off the rails. When a child picks up on a parent's pain, they absorb that pain to keep up the belief that their parents are all-powerful and unbreakable. If the child thinks a parent is flawed, he or she will feel unsafe; pain is

preferable to feeling fearful. When they're very young, they believe they're the culprit who makes Mommy or Daddy upset. Later, they believe Mommy and Daddy are the culprits who upset everyone.

The Greatest Things You Can Teach Your Child (Things I Wish I'd Known Before)

Long-term studies show that a child's future success can be boosted by teaching them emotional and social skills. These skills are the most important . . . outside of learning Mandarin.

1. Teach your kids to resist instant gratification. If you can teach them to control their impulses, it will benefit them throughout their lives. Research was done on three-year-olds and their progress followed for thirty years. It was found that those who developed self-control had better physical health, were better educated and attained a higher level of achievement. Children without the skills to delay reward were more prone to drug use, had more chance of engaging in criminality and had a lower standard of living.

2. Teach your kids to pay attention. If you teach them to focus, they'll be more able to filter distractions, especially in this age of information, where trillions of bits are incoming each moment. When you pay attention, stress levels lower, and that's about as good as it gets so far as passing something useful down to your kids goes.

3. Teach your kids empathy, to try and imagine what it's like to be in someone else's shoes; to feel and think how another person does. This nips bigotry in the bud.

How to Develop Your Child's Brain (Things I Wish I'd Known Before)

I like to have physical evidence of what's happening in the brain to validate any advice I give or take on how to make your kid a better human being. If I can't see it, I'm not buying it. I'm not going to crack open my child's brain, but I like to get a rough idea of the interior landscape.

1. Link up the left and right sides of the brain. Every human comes equipped with a left and right side of their brain. The left side helps them to organize and make sense of their thoughts; the right allows them to pick up the vibes of a situation; get the bigger picture and what's going on below the words, deduced through tone, posture, facial expressions and insinuation.

   The left side can make sense and put into words what the right, emotional, side is experiencing. If a child is swamped by emotions, unless someone helps them, they'll just wade deeper and deeper into self-doubt and chaos. So, when your child is floundering mentally, it's you who'll help them find the words and make sense of the feelings. Like a translator for a hurricane.

   To tune into your kid's right brain, link up with them by using your own right brain, which picks up clues about their feelings through facial expressions, movements and tone of voice (specialities of the right brain). This is called right-to-right attunement. Eighty per cent of all communication is done by these below-the-radar clues, rather than speech.

   This right-to-right attunement is when the parent connects biologically with the child by mirroring emotions back to them, making them feel understood. Then, the parent can redirect the feelings from the child's right brain to their left (by using their own left brain) to help the child figure out a logical explanation and put their feelings into words. When the right and left brain hook up, the feeling of helplessness and confusion passes and balance returns.

2. Link up the older to the newer brain. Your child has a limbic brain (as you do), which is perfect for those quick, fast responses we need to survive. What they don't have at birth is the more evolved, higher prefrontal cortex. Your job as a parent or caregiver is to help your child grow that part, which isn't completely developed until they reach their mid-twenties. So not only are you helping to integrate the right with the left side of the brain, you're also aiming to combine and coordinate the top with the bottom to create a well-balanced brain. All this integration has to be taught by the parent for the child to learn how to make better decisions, gain self-control, hold off from instant gratification and regulate their emotions. You mould your kid's destiny.

When the Going Gets Tough (Things I Wish I'd Known Before)

Your kid will have tantrums, just as sure as they'll grow toenails. What you might not know is there are two types of tantrums, one an eruption from the higher brain, the other from the lower.

The higher one is when your kid throws a fit to get their own way, acting out a scene but aware they're acting. They know exactly how to manipulate you to get what they want. If you give in, you're making a big mistake. This is the time to send in the troops and set the boundaries by saying you understand the frustration but rules are rules and, if they don't stop misbehaving, they'll be locked away in a tower and given a poisoned apple . . . or you could decide to be less crazy and just tell them to go to their room.

The lower-brain tantrum needs a completely different approach. If your kid has gone into 'full hissy', awash in cortisol, all reason has left the building. Now, all you're left with is a baby rampaging bull, so either physically move them away from people they might disturb or, if no one's around, remove yourself. When they've calmed down, choose your moment to discuss what happened. Without being demonstrative, ask them to describe the stages that led up to the outburst and what you could do the next time to help. Let them try to figure it out themselves so they become adept at problem solving. If they refuse to discuss it, just inform them in a calm tone of voice that if they disrespect someone or throw things, that behaviour is unacceptable. Stay consistent in your reactions. If you fluctuate, the bull will stampede you.

Here are a few tips on how to reconnect when your kid's emotions have gone into hyper-drive.

1. Empathize

The next time your kid is in crisis, say because their turtle died, rather than saying, 'Pull yourself together, for God's sake, turtles die,' you could help them express themselves by mentioning that they look sad, or tell them you can imagine how bad they feel because you had a turtle once too. (Even if you didn't have a turtle, just say it.)

2. Physical contact

If your kid is hysterical, don't try to talk sense into them – there is no sense in them. Try to hug them (not hard, even though you may want to strangle them). Even with a light touch, cortisol levels reduce, oxytocin levels increase. If they are in hyper hissy-fitting mode, do not hug them, it will drive them even more crazy.

3. **Make faces**

Use non-verbal signals such as empathetic facial expressions, a soft tone of voice and listen non-judgementally. But make sure you're not unconsciously trying to make things better by making a 'boo hoo' sad-clown face. They will only hate you for that one.

4. **Physical distraction**

If the hysteria continues, point out something novel or funny that might make them switch their attention. You could suggest a piggyback ride to gallop them out of their misery.

5. **Storytelling**

Later, when the storm has passed, maybe ask your child to tell you a story about what just happened. You can show empathy by interjecting (but not too much) phrases like 'I wonder how that felt for you . . .' Let them fill in the blank how-it-was-for-them. The fear isn't so overwhelming once they've named the pain.

6. **Draw it**

If they don't want to talk about it, maybe suggest they draw or paint a picture of how they felt or are feeling. The unconscious has a way of sneaking out when you have a crayon in your hand.

7. **Let them blow off steam**

To release the fuse, you could suggest jumping jacks, going for a run, playing a ball game, cycling, throwing eggs . . .

You change the emotional state by changing the physical state.

8. **Lay off**

   Sometimes, they need to be alone and really don't want to talk. Check that out as an option if nothing's working. The instinct is always to make things better but, sometimes, it's better to let them figure things out.

Teachers

I'd like to put in my two cents about how I think teachers should teach. Children's brains are like landmines that can go off later in life if they're put under too much pressure. It starts in school, where they're forced to achieve even if they have absolutely no interest in a subject. Intelligence at school is determined by whether you have the skills to memorize facts and spew them out during exams. I call it 'mental vomiting'. And some of the teachers I had, rather than figuring out how to make something compelling by using a little imagination, ended up making me hate the subject they were teaching.

My Story

I was fascinated by volcanoes in sixth grade. Mr Viveric, a great-looking teacher with a great personality, told us to create our own volcano. He gave us no instructions on how to build one and set no limits. I spent weeks on it, obsessed, and finally brought in an eight-foot volcano made of papier mâché. I don't know how I knew how to make what would today be called a bomb, but I did. When I lit the match, my volcano not only erupted but started to burn the ceiling down. I got expelled (again).

If I had been rewarded for knowing not only how to build a bomb but also how to defuse it, I'm sure, later in life, I would have been asked to join the Special Forces but, thanks to the narrow-minded educational system, I am now a comedian.

So much could be taught through games rather than this thing we call 'homework'. Why can't teachers make learning about things like volcanoes fun? What's more fun than volcanoes? If you are forced to write a paper on their geological origins and get an A, you'll probably lose your mojo. Plato knew all this back in BC. Why can't we get it into our heads? He said, 'Study forced on the mind will not abide there . . . Train your children in their studies not by compulsion but by games.'

How to Teach Kids

1. **Games**

Most parents automatically know how to excite their babies, when, for example, they play peekaboo. The baby gets a jolt of an 'aha!' moment from the surprise of seeing the mother's face pop out, so learning something new becomes associated with something thrilling. The excitement of new information comes with a hit of dopamine as a reward. Once you've got them motivated, they'll want to learn more and more to get another hit of that dopamine.

A note here: I don't think it's an addiction when you want to learn more. I don't know anyone who's an addict to knowledge and, if they are, they're helping the world. I say, go ahead, give them as many hits of information as they want. If the teacher or parent gets the cocktail of dopamine, serotonin, adrenaline and endorphins right, a child will be like a magnet for incoming knowledge.

2. Change the environment and method of teaching

Investigate which learning environments and study methods work best for your child. We aren't all proficient at learning while we're sitting stationary. Remember: our forefathers learned about nature, food, weather, architecture and survival by being mobile. We all have a unique way of memorizing and taking in facts. My daughter could memorize facts if she sang them aloud. So I let her sing. Sadly, they didn't let her sing during her A levels, so she didn't do so well. Another kid may learn better when they walk around. I do my best work in the shower or in bed.

I know these individual methods would cause havoc but maybe someone, someday, could think about dividing a classroom between singers and walkers. If it's too disruptive, at least experiment at home by encouraging your child to find their own methods. This is especially relevant for kids who have learning difficulties. Just help them find the key to unlock their interest and watch what happens.

3. Tribal classrooms

We are at our happiest when we work in tribes; that's how we thrived and survived. Let's bring a little of that into the twenty-first century. An effective way to teach kids is to encourage them to work together as a team, and not to pick out the strong from the weak. They'll work harder and help each other when they know they're working for the good of the gang. The goal is for the kids who are lagging behind to be taught by the smarter ones.

4. Creative thinking

Schools should give high marks for thinking out of the box. The teacher should ask why the students answered the questions the way they did and, if it's imaginative and original, give them a high grade. If someone keeps getting disapproval because of a bad grade, they shut down and no longer feel safe to explore other options. If you repeatedly tell someone they're not doing well, they will underperform. Also, gold stars should be given to young kids for waiting their turn, sharing their toys and demonstrating that all-important emotional intelligence.

5. **Teach them how to fail**
   They are going to fail at some point in their lives, so get them ready. Most of the kids who got the straight A's in my high school are now out of their minds, probably because they were pushed to the limit at the age of two. I went back to find out what happened to the Prom Queen of my high school – cokehead. I swear to God I didn't smirk. (Maybe a little.)

My Story

When I was in high school, very few teachers knew how to capture my interest. During classes, I was so bored I'd doodle on the walls, using mustard as my medium. If only a teacher had noticed this early talent rather than getting me expelled, I could have been another Jackson Pollock. Sometimes they should focus on what you're good at rather than what you're bad at. Once you can pay attention to something, even if it's looking at bugs on the windowsill, you should be applauded because the ability to pay attention is one of the greatest skills of all, so if you're rewarded for being able to hold focus, rather than on your results, you should get a gold star.

6. 'Difficult' kids

Those who are considered difficult children usually come from chaotic backgrounds; they're in so much turmoil, they don't have up-to-scratch social skills or high levels of concentration. The memory doesn't function well if a person is under stress, so learning is difficult. What looks like bad behaviour on the surface is usually fear and anxiety, along with an inability to regulate emotions. Children learn best when they feel protected and connected. You can compensate for a lack of nurturing in early life by gaining the child's trust. This cools down their chaotic mind and opens them up to information, because now they feel safe. I said earlier that all of us have a leaning towards negative thinking; if you give five positive statements and one is slightly critical, it's the critical one that's remembered. With difficult kids, aim to give them a lot more positive feedback than you might otherwise. Find out which areas they might be creative in and let them teach you.

A Reality Check

I want to flag up a few facts. The average unemployment rate after leaving university is about 15 per cent, and as high as 22 per cent in some areas. I think it might be an idea for kids to learn occupational skills that will ensure they get a job later: plumbers, electricians, carpenters. Then go to places like Notting Hill Gate or Hampstead and name their price. Everyone there is desperate for good tradesmen. I'm sure some time in the near future, plumbers will be the new bankers.

Another fact: 65 per cent of eight-year-olds will work in jobs that don't as yet exist, so why hot house them for something that will be redundant? Why are we pushing kids to learn subjects they probably won't need in their future? The skills of the future will probably be about being creative and thinking out of the box. Those who can do those 'soft' skills won't be replaced by a machine. Imagination may become a valued commodity. 'Thinkers out of the box' will go to Harvard or Yale.

A machine won't know how to empathize, how to read what another person is thinking. Computers can identify a face but not what's going on underneath it. Empathy becomes the new literacy. These days, those with people skills are the rising stars because, in this global world, you have to know how other people think and feel.

Businesses look for leaders with 'people skills' because, no matter how intelligent someone is, if they can't relate to other humans, they're not going to float and neither is the company they work for. Creating trust should be as highly rated as hitting money targets. Cooperation, curiosity and good communication skills are the 'it' girls.

A Short Note on Teenagers (Things I Wish I'd Known Before)

I know I haven't mentioned teenagers up until this point. How to deal with them requires writing another book dedicated completely to the topic. I may do that in the future but, in the meantime, let me tell you that most of their abuse to you is nothing personal, it's because, inside their bodies, the equivalent of Hurricane Katrina, the San Francisco Earthquake, Pompeii, the Yangtze River flood and the Evado del Ruiz volcano eruption is happening. This is the time in their lives where their biology is shifting like tectonic plates, moving them from childhood to adolescence. Even though they now have bigger protrusions, they still partly have the brain of a baby, but they don't want to be treated like a baby and this is why everything you do is an irritant, from the way you breathe to your very existence. Boys usually cut off from their parents by distancing themselves i.e. locking themselves in their room; girls usually go into combat mode and let you know how irritating and passé you are. If you wait a few years and you survive, I promise they will love you again.

If you tell them, 'No,' don't add on long explanations, because they are geniuses at manipulating you with every ploy on earth. Here are some of the more common comebacks for when you say, 'No,' to something as simple as, 'No, I don't want you to stay out until four in the morning' to your fourteen-year-old:

You hate me and want to ruin my life.
You can't tell me what to do.
You never let me do anything, you're such a bitch.
You're just jealous cause you're a loser.
Asking you, 'Why?' over two thousand times an hour.
Everyone else's parents let them do/have/smoke it.

Some rules:

1. Don't just say, 'No.' Always present a simple case for why you said it, to show you've thought it out and, whatever you do, don't elaborate.

2. If you get into the cycle of reasons for your 'No', your child's retaliation will be endless, leaving you drained. You'll be like an old, dead, dried corn husk and will eventually change the 'No' to 'Okay, whatever you want.'

3. If they find out they can wear you down, that there's a teeny chink in the armour, they will gnaw away at you until you're swallowed.

4. If their argument is reasonable, you can change your mind to show them that you do listen and are at least partially human.

5. Choose your battles. Too many 'No's may drive them away, shut them down or they'll do things behind your back. Decide which are the absolute 'No's and when you can throw them a 'Yes' bone.

6. If the argument gets too heated, stop and wait to respond when you're feeling calmer. Even if they've gone too far and insulted you, wait until you've cooled down to discuss why it will never be acceptable to abuse you.

7. If you do lose it, you lose it; it's not going to harm them if they see your anger when you're really angry. If you've gone too far in expressing it, wait until you've simmered down and then, without expecting a reply, say, 'Sorry.'

The Monk, the Neuroscientist and Me

Ruby: Does a neuroscientist bring up a kid differently from the rest of the world?

Neuroscientist: We have a tendency to do experiments on our own children – it's a captive audience. For example, there's a technique that developmental neuroscientists use with an electronic pacifier. By recording how often the baby is sucking, you can tell whether they are paying attention to what you're showing them. When they're alert and interested, they suck more, and it shows us about their early cognition.

Ruby: You're like Dr Frankenstein. And if they don't pay attention, do they get an electric shock? Is that how you train them to be smarter, rather than just be a baby? You might electrocute them but at least they'll go to a good university.

Neuroscientist: I wanted to do those kinds of experiments with my son, Kirin. It was the first time I had unlimited access to a baby. My wife didn't really go for it. But we did try the things that lots of scientists try with their kids. We bought the big black-and-white mobiles for the crib, for example, because the early visual cortex can be patterned by high-contrast shapes.

Ruby: Did it help?

Neuroscientist: I don't know. I shoved a lot of black-and-white shapes in his face when he was little.

Ruby: Yeah, I got my kids mobiles, too. I don't think they became geniuses, as far as I know.

Neuroscientist: Were your mobiles black and white?

Ruby: No.

Neuroscientist: Exactly.

Ruby: But why would that make him smart?

Neuroscientist: There's stuff that you do as a neuroscientist because you know something about the brain and you think you can intentionally improve brain function. I feel that about my own brain, but even more so about my son's because it's literally growing in front of my eyes. We have all these theories about what should work and what shouldn't work. I mean, none of it really pans out.

Ruby: Do you sometimes feel like you're growing his brain in a lab?

Neuroscientist: Definitely.

Ruby: So, how are you growing your son's brain?

Neuroscientist: Well, my wife stopped me doing a lot of the stranger things. She says we should love him, care for him and talk to him.

Ruby: What a crazy thing to do.

Neuroscientist: It's crazy, I know.

Ruby: If she had let you loose, what else would you have done to your son?

Neuroscientist: Oh, there would have been a lot more experiments.

Ruby: Like what?

Neuroscientist: Well, I'm very interested in the relationship between finger counting and numbers. I wonder what would have happened if I had bandaged up one of Kirin's hands so he could only count on the one hand? Would it change his mental number line? Would it help or hurt his acquisition of numbers? That sort of thing.

Ruby: And what would that have done?

Neuroscientist: I don't know. I think it would have been interesting.

Ruby: What else?

Neuroscientist: There's a great experiment where you have Bert and Ernie puppets.

Ruby: Who are Bert and Ernie?

Neuroscientist: They're the gay couple from *Sesame Street*.

Ruby: They're puppets, I hope?

Neuroscientist: Yeah, they're not openly gay. So, you've got multiple Berts and multiple Ernies. And then you've got a little stage. The curtain closes, you put a Bert and an Ernie behind the curtains on the stage. Curtain opens, there's Bert and Ernie and they're talking to each other. Curtain closes, then opens again. Now, instead of Bert and Ernie, there are two Berts or two Ernies, so a Bert has become an Ernie, or vice versa. Then you measure how many times the child sucks on the electronic pacifier.

Ruby: What does this show? This is so frightening.

Neuroscientist: It shows whether babies are most sensitive to changes in colour, identity or number. So, instead of having a Bert change to an Ernie, you could have a Bert and two Ernies.

Ruby: And they would suck on the electric thing?

Neuroscientist: As they become habituated to Bert and Ernie, they suck less. So, if the sucking rate goes up after the change, it means the baby noticed the change.

Ruby: And does this teach them to be smarter?

Neuroscientist: Well, no, but it's interesting.

Ruby: For you?

Neuroscientist: Yeah, I guess it's just interesting for me.

Ruby: Great. Give me some other tips on childrearing.

Neuroscientist: I tried to teach him Latin.

Ruby: You did?

Neuroscientist: Yes, early. Around age three.

Monk: I'm phoning Child-line.

Neuroscientist: No! I just tried to teach him some Latin.

Monk: How did you do that? With flash cards?

Neuroscientist: I just started naming things in Latin. I would make him use Latin names for things. It didn't last long.

Monk: What else did you do?

Neuroscientist: You know that age when kids start to name body parts, like 'Knee!' 'Elbow!' I taught him the medical names for body parts. He knows where his scapula is and that he's got a uvula and that when he swallows a bite of food it's a bolus. Oh, and I taught him some quantum physics when he was five. Hours of quantum physics.

Monk: That's just insane, but tell us more.

Neuroscientist: It started out innocently enough. Someone gave us a book of chemical elements or something and it had the periodic table in the front cover. I love the periodic table, I think it's a beautiful and amazing thing. So, I started taking him through it, and I said, 'Look, this is copper, this is what it looks and feels like. This is sodium. It's also a metal but it looks and feels really different.' And at first he was excited about elements, he would go around the house and he'd say, 'Oh, this is glass. It's made of glass.' And I would say, 'Yeah, now let's talk about the atomic structure of glass. Why do silica compounds behave in such interesting ways?' That led us into electron shells and why subatomic energy is quantum.

Ruby: How old was he?

Neuroscientist: Four or five. I don't know if he remembers all that.

Monk: Right.

Ruby: But just explain to me what would be your hope when you cram his brain? He knows Latin. He knows quantum physics. He knows the reason for the universe. And now, imagine you're dying, what was it that you wanted him to have in life?

Neuroscientist: It's not that I think he needs to have any of this information to succeed in life, I just want to take advantage of the fact that he's very young and his brain is like a sponge at that age. Babies start out with trillions of connections between their neurons, more than they will ever have in their lives. This is the best time for them to learn a language, a musical instrument, or anything. I want to give him everything I can in that window.

Ruby: So, what's your deadline? You want to stuff it in by what age?

Neuroscientist: Five or six. After that, of course, kids can still learn lots but it just takes more effort.

Ruby: Right. And then, were you hands off?

Neuroscientist: I've been a little bit more hands off since then, but that's probably because of my wife's intervention. She just wants him to be happy and to be able to do nothing sometimes so she stopped leaving me alone with him so much.

Monk: That was a good move. She probably has a child psychologist on speed dial.

Ruby: How old is he now?

Neuroscientist: He's eight.

Ruby: What does your wife want for him?

Neuroscientist: She wants him to be well adjusted. It's not in the Indian tradition to go for well adjusted.

Monk: We've just all got to be doctors.

Neuroscientist: Well, not everybody has to be a doctor, but everybody should go to medical school.

Ruby: Wow. Would him being well adjusted be upsetting for you?

Neuroscientist: I mean, I can see the value in well adjusted, it's just not what I'm bringing to the parenting.

Ruby: There must be a point where they go, 'I hate my parents.'

Neuroscientist: I hope that won't happen!

Ruby: What if he didn't have the kind of brain that could learn that stuff? They're going to hate you if you try to enforce that information. My daughter would have sued me at age two.

Monk: They want to play on the Xbox and you're teaching them classical Greek.

Neuroscientist: It's a risk. But I also think you've got to be the parent you can be. The parent that I can be is someone who likes Latin and physics. That's just the parent that I am. I can't really be the parent that goes to the bouncy castle and wants to read *The Very Hungry Caterpillar* all the time.

Monk: But you'd do that for a friend. What if I said, 'Ash, I really want to go and play on a bouncy castle.' Would you come with me?

Neuroscientist: I would once. We would hang out less after that.

Ruby: Really? So, you'd take Thubten to a bouncy castle, but only once?

Neuroscientist: I would once.

Monk: But not again.

Neuroscientist: Not again. But Thubten on a bouncy castle would be a marvellous thing to behold. The saffron robes flowing up in the air, the smiling face, the gleaming bald head . . .

Ruby: I'd pay to see that. I wonder how much stress is too much when you're raising a kid?

Monk: I think the main thing is to make sure we're not passing our stress on to our kids. But the trick is also knowing when to push, in the right way.

Neuroscientist: That's the hardest thing to work out. It works well when you push and then let up. Like when I'm working on a difficult problem with lab data, I immerse myself in that problem and work hard. But, at some point I have a shower or go for a walk and, many times, the answer just pops into my head. I call it a shower epiphany. It happens when the stress levels drop, then the solution pops into your head fully formed. But it won't pop into your head if you don't do the hard work first.

Ruby: I agree with you. If you don't push in the beginning, you won't ever become proficient at anything. In the beginning, when you learn something, it's always agony and, later, it's smooth sailing. I wept when my piano teacher put the lid down on my fingers when I screwed up *Für Elise*. Eventually, I loved playing piano, even though my fingers were broken. But I don't always know when to back off and take a shower. Sometimes, I never do. How can I teach my kids if I don't know when to back off myself?

Monk: I think the more you get to know your own mind and learn how to regulate your stress, the more you can do the same for your kids. If you're pushing them because you're afraid they'll fail, they won't learn anything. It can't be about you. You have to tune into the individual child's innate talents and nurture those. There's no sense pushing what they have absolutely no interest in. My father used to whack me to learn maths, and now I go blank when faced with numbers.

Ruby: Are you kidding?

Monk: No. I got whacked quite a bit. It was a heavy time.

Ruby: Was he happy when you became a monk?

Monk: Actually, he was also a monk for a short while before he met my mother. When he first saw me as a monk he was so proud of me that he lay prostrate in front of me on the floor. It happened at Heathrow Airport. It was quite a moment.

Ruby: Did you ever forgive your father?

Monk: What changed it for me was when I found out about his childhood. He was brought up by a very austere mother. One day when he was young, he came home from school really excited because he had spent all day carving a wooden spoon for her. He handed it to his mother and she stared at him and said, 'You encumber the earth.' It helped me understand him better and understand his pain. We're friends now. He's a really interesting guy.

Ruby: Thubten, how would you raise a kid?

Monk: Well, I think I'd get them to talk about how they feel and I would try to be non-judgemental. And, like I said, I'd try not to dump my emotions on them.

Ruby: So, let's say I was angry and I said, 'I want to bite my doll's head off and stuff it down Lumpi's throat' (he's my dog). You'd go, 'Okay, let's talk about that?'

Monk: Not right away, but in a calmer moment later I'd say, with kindness, 'Why would you want to do that? What were you feeling?' You can have intelligent conversations with children about how they feel, trying to stay curious instead of blaming, and not patronizing them.

Ruby: Thubten, you should have a child. The world would be a better place.

Monk: I really can't do that, but thanks for the thought.

You'll find the relevant mindfulness exercises for bringing up kids in Chapter 11.

9
Addiction

Many of us feel we're on the brink of a new Ice Age, an economic plummet or just biding our time before the North Koreans twitch their trigger finger. All this may happen (hopefully, not until after my book is published), but the main threat to our longevity is our addictions. We know how to deal with limited resources; we don't with unlimited resources. In the old days, if we were hungry for a snack, we could just grab a banana even though there was a limited amount but, now, in this click-of-a-finger society, we can get immediate gratification. One click and you can order in porn, food, clothes, cars, jewellery, a husband . . . and I haven't even got to the Dark Net yet. We did not inherit handbrakes for when too much is too much. Up until now, in the West, you could never earn too much but, now, with the newly minted billionaires, it's unfathomable to imagine how they deal with their lives when they can have anything in unlimited amounts. Show me a healthy billionaire who's having a whale of a time and I'll eat my socks.

Novelty

What kept us progressing throughout our evolution was a quest for novelty. This is what motivated us to invent the next 'big thing', from spears to missiles. Studies of DNA show that Neanderthals carried a gene called DRD4-7R forty thousand years ago. That gene is associated with risk-taking and sensation-seeking – always seeking, but rarely satisfied. A similar gene exists in 10 per cent of the current population, and this 10 per cent are more likely to be addicts, if not the greatest daredevils (see extreme sports addicts). I'm not saying we can throw caution to the wind, stay addicted to our substance of choice and blame it on our Neanderthal genes but, if we're aware of our propensity towards addiction, we stand more of a chance of being able to do something about it.

In ancient Greece, the word 'addiction' meant 'those not entitled to rights', in other words, slaves. In a way, that's a pretty good description, because when you're addicted you have no freedom, you're always enslaved to your drug of choice. I'm somehow sure that no one gets addicted to brushing their teeth, but I'm probably wrong.

Digital Hits

In the past, chemists created dangerously addictive substances; today, entrepreneurs get us addicted to devices. They get rich by finding some new way to tickle our sweet spot to keep us hooked. They know they need to keep the hits coming to ensure no one ever gets bored; the consumer always needs higher doses. These digital-media honchos don't accidentally trip on to the 'next big thing', they run thousands of tests to learn exactly what gives the user the biggest buzz and how long to hold back before giving them the next one to get the level of craving just right. The toys are engineered to be irresistible to all who touch them. Their creators research what stokes you up to the max, using specific background colours, audio sounds, animations and music.

Before the internet, we were manipulated by advertisers (see *Mad Men*), who knew very well what would draw our eye to their particular product. In the sixties, *The Hidden Persuaders* was published, which discussed how we are suckered into choosing one product over another, how the admen made sure we reached for the detergent that made our whites 'whiter than white' rather than any of the other detergents that did exactly the same thing.

The difference now is that, with the advent of digital advertising, we're being manipulated every second, day and night. We are living in an 'attention economy', where admen/women make a living by knowing how to turn our most precious commodity – our attention – into hard cash. Their metric for our attention is referred to as 'number of eyeballs'. Everywhere you look, they're competing to get our focus: on the street, onscreen and on public transport. In the past, we were at the mercy of billboards to lure us in but, when the internet went mobile, addictions soared. In the sixties, people got hooked on cigarettes, alcohol and drugs. By 2010, people were hooked on Facebook, Instagram, porn, Twitter, Grindr, Tinder, online shopping, box-set bingeing, and the list goes on.

Each month, each person spends almost one hundred hours texting, gaming, emailing, reading online articles, checking bank balances, etc. That adds up to about to eleven years over the average lifetime. Forty-one per cent of the population have suffered from at least one behavioural addiction over the past twelve months. Today, most people spend on average three hours a day on their phone: that's a quarter of your waking life plugged in. There's even a word that describes the fear of being mobile-less: 'nomophobia'. Eighty per cent of teens check their phones at least once an hour. Most of them have learned to never look up from their screen or tablet and live their lives in a permanent state of distraction.

That Good-time Drug: Dopamine

Both substance (hard drugs) and behavioural (sex, shopping, etc.) addiction involve the release of dopamine produced in the ventral tegmental area (for you neuroscience nuts); it's then sent to receptors in the brain, generating a rush of pleasure. We need a small dose of dopamine to initiate almost every action we do. Even drinking a glass of water has a reward in it: to stop the craving of thirst. It was our need to survive that created this reward system. When we came upon a source of food to satisfy our hunger, we'd remember where it was so we could find it again. The dopamine rush has three elements: trigger, behaviour and reward.

The problem today is that our reward-based learning system goes into overdrive with no brakes because, with enough money, those sources of pleasure never need to run dry. When you have constant dopamine rushes, the brain decides to stop the gushing. Now, you don't get the same buzz as you did before, so you need to up the dopamine to get the same kick and, if you don't, you're left with the craving. Eventually, you'll need great waterfalls of dopamine just to feel normal and may have to take up even more harmful addictions, for example, cocaine, to fill that void of need. When you get a craving, an area of the brain called the cingulate is activated. fMRI scans show that this region quietens down in those who practise mindfulness. People who practise have learned not to latch on to the thoughts about the craving but instead regard the sensations as simply sensations.

You can't depend on your thinking, because it knows how to justify your urges, as in, 'I'll only have a little bit, then I'll give up,' or 'I'm not addicted, I just like smoking,' or 'I'm snorting this because my mother screwed me up.' But if you learn to sense that urge in your body early enough, you'll gain the ability to pause and make a choice to take or not to take another dose of whatever it is before it swells into the full Billie Holiday story.

Some people can do certain drugs or indulge in certain behaviours and not get addicted but, if you're highly anxious or depressed, you stand more chance of getting hooked, because you learn that whichever substance or activity you're addicted to lessens the pain. If you need to keep doing or taking something to get emotional relief, after a while you'll feel you can't live without it. Then you have become, officially, an addict.

How to Get Over Addictions

What can you do about it? AA, NA – all the A's are a brilliant answer to addictions and help millions of people. I think part of what makes them so successful is that they provide a community, and that resonates with our primitive instinct to want to be accepted and included. This sort of community brings out the best in humans because it's a space where there is no judgement or social hierarchy. Each man or woman is considered equal and all are cared for, no matter who they are.

What keeps many addicts hooked is that feeling of inclusion; hanging out with other addicts. AA and NA give the user a replacement tribe, but now the members help each other to have a better life. The community has its own rules and boundaries, which make the members feel safe. Those who go to AA or NA hold back on instant gratification for the good of the group, and this is a throwback to what made an ancient community successful. If individuals didn't hold back on food back then, the tribe would starve.

An addict can recover, but they must be willing to give it up. Neuroplasticity can come to the rescue, because none of us has to be stuck in a habit, we can take control and decide to change our brains. The difficult part is that it takes time and discipline. Whether you practise mindfulness, go to meetings or see a counsellor, the key is to become aware of your thoughts and feelings. Then, you have a choice to pull out or stay sucked in.

The Monk, the Neuroscientist and Me

Ruby: Ash, do you think there's an addict's brain?

Neuroscientist: No, there is no such thing as an addict's brain; you're not born with it. There's no way to tell, looking at brains, or genes, or anything else, who's definitely going to become an addict and who won't. Certain things make addiction more likely, like childhood trauma and parents who are addicts and, obviously, access to drugs makes a big difference. But even these risks are not absolute. Some people at low risk will still become addicts, and some people at high risk will not and, so far, we don't know why.

Ruby: What's the difference between a habit and an addiction? I mean, is shopping an addiction?

Neuroscientist: I'd say that a habit is a behaviour with some choice but an addiction is a compulsion with much less choice. If you don't feed a habit, it may make you feel uneasy, but people will turn their lives upside down to feed an addiction. And withdrawal from addiction can make you physically ill. One of the reasons people keep drinking is to relieve them of that jittery withdrawal. Most alcoholics don't drink for a buzz, they drink to feel normal.

Monk: I want some chocolate. I know you've got some in here. (Starts going through the cupboards.)

Ruby: Ash, is Thubten an addict?

Neuroscientist: He's out on the sidewalk selling your TV for chocolate right now. You tell me.

Monk: I struggled with addiction before I became a monk. Then, when I joined the monastery, I even got addicted to meditation. I was doing it to get a buzz, to feel high, like drinking a triple espresso. Instead, after a while, I started to get depressed, carrying a heavy feeling around all the time. I went to my teacher and told him that meditation was making me depressed. He said that it wasn't the meditation, it was me. He said, 'You're a junkie. You're trying to get high through your meditation, but grasping can never be fulfilled, you'll always be looking for more. The sadness you feel is disappointment.' It was a breakthrough because it changed my attitude to the practice.

Neuroscientist: I think that's a profound insight. You're looking for that hit of pleasure from meditation and, actually, you're practising noticing the absence of pleasure.

Ruby: I'll tell you what gives me a hit. I'm addicted to Netflix. After episode one, I'm hooked, then I spend the rest of the night chasing the dragon, eating my way through box sets. This is how nuts it is. I was invited to go to the cast party for *The Crown* because I know the writer. I didn't go because I was so hooked, I had to stay home to watch episode six.

Neuroscientist: Yeah, binge watching. It's like that experiment they did with the rats, pressing levers to get more drugs.

Ruby: Rats watch Netflix?

Neuroscientist: Yeah, I heard they love *Breaking Bad*. No, I'm talking about those experiments where rats pushed levers to get access to cocaine. The story is that the rats gave up food, sleep and exercise and just kept pressing the cocaine lever over and over. Eventually, many of them died of starvation. That's exactly what it looks like to me when people keep pressing the Netflix button to get the next episode.

Ruby: In your eyes, we're all rats. Thubten, do you watch Netflix?

Monk: Have you guys seen *GLOW*? I mean, I'm not usually into women's wrestling, but this is really about the human condition.

Ruby: The human condition? Yeah, in Lycra and tearing each other's hair out.

Neuroscientist: I think that's a fair assessment of the human condition.

Monk: Actually, I don't really watch television that much, because I'm really busy.

Ruby: Isn't being busy also an addiction?

Monk: I definitely get a lot of joy from working hard and helping people, but it's not a toxic 'buzz'. I guess if I were getting paid, I might get addicted to making more and more money.

Ruby: Money seems to be the most addictive of them all. The ultimate buzz.

Neuroscientist: Yeah, money is like mainlining drugs. It's like we designed money specifically to give us a dopamine dependence. We get the hit of pleasure, but it doesn't last, we always want more. Big banks use that infinite craving to drive people to do practically suicidal levels of work. It's a lot like those rats with the cocaine levers.

Ruby: I heard that at a big investment bank – I won't say the name, but the first word starts with a G and second one with S – when they hire new recruits, they give them the same test psychologists give people to see if they're psychopaths. If the potential candidate gets high scores on the psychopath test, they get hired. It indicates they'd kill to get a deal. If we revere maniacs now, what do you think is going to happen in the future?

Neuroscientist: We're committing more and more to a system that rewards short-term dopamine hits over long-term happiness. Your brain produces dopamine as a reward for taking risks, not for making sound long-term decisions.

Monk: People get addicted to the dopamine. I think it's interesting that they don't get addicted to organic kale, but they do to sugary drinks and so on. It's all about having our senses ramped up, a kind of buzz. Then we end up swinging between the highs and the lows, and contentment seems boring. I was scared of becoming 'bland'. When I first joined the monastery, I phoned my family in a panic, saying, 'What if I end up like an automaton, with no feelings?' I think I was worried that meditation would take away any sense of 'pizzazz', and that it would all become very grey. But, actually, I now feel that happiness is all about a stable sense of inner joy. When you don't need a high, you can feel great.

Ruby: Yes, part of being happy is not having to constantly chase it.

Neuroscientist: Exactly, but commercial businesses are all about hyping up the chase. Right now, Google knows from your search history that you've been thinking about buying shoes, so they'll put a shoe ad next to your email. You think it won't affect you, just like all addicts think they're the exception and that they won't get addicted. But Google knows it will get you, and they know how to do it. They know when you're going to be most sensitive to that ad, they'll tailor the ads right up to what you were thinking about in the last hour. These are manipulative technologies and they're designed to create and feed addictions.

Ruby: When you go shopping, do you think there is some kind of Big Brother manipulating you to buy stuff?

Neuroscientist: Yes, definitely. Shops use lighting, smells, music and layout to create fantasy experiences. They give you beautiful fountains and cheap food so you feel like you're on holiday and you'll be more frivolous with your money. Every piece of the retail experience is designed to make you give in to your compulsions, to make an impulse purchase without thinking about it too much. You've opened your wallet even before you've decided to buy anything.

Ruby: How do I avoid all that? Do I just have to stay in my house?

Neuroscientist: That's not going to help because, now, the store comes to you. It's on your computer even when you're just checking your email; you're shopping before you even made a decision to shop. Normally, one of the defences against addiction is not putting yourself in situations which might feed your addiction but, with the internet, you have less control over what comes into your life.

Ruby: If you're an addict, what do you suggest doing?

Neuroscientist: We've known for a while now that changing your environment is one of the most effective things you can do to combat addiction. The best evidence for that view came from US soldiers after the Vietnam War. A lot of them became addicted to heroin but, when they came home, most were able to stop using. Heroin is a very difficult addiction to kick, so that's a big success story. Part of that was because the army provided good support and monitoring, but the biggest factor was that the environment changed. Back in their homes in America, it was much harder to get access to drugs, it was less acceptable to use and, for the most part, the psychological trauma of the war was over. That's why it's important to focus on social changes, not only on biological ones.

Monk: Ash, I agree with that focus on changing environments. I've taught mindfulness in addiction clinics and I suggest to addicts that they change their circle of friends or maybe rearrange their house. They could change their living room into a kitchen or the bedroom into a living room. It sounds simplistic but, if you rearrange things, it breaks up your routine and can prevent you from falling into old habits. Shake up your environment and it can give you a fresh start and a chance to move forward.

Ruby: Seriously? You're going to tell a heroin addict to get a new kitchen and that's going to work?

Monk: We don't stop there. Then we talk about the inner environment. I talk about how addictions are like scratching a wound. The more you scratch, the more it itches or gets infected. If you can just experience the itch and hold back on the scratching, it can start to heal. That's where mindfulness comes in.

Neuroscientist: That's a great analogy. Wounds release a molecule called histamine, which helps the healing but also causes the itch. Scratching releases even more histamine, so you get even more itching. If you can put off scratching, if you put a slight gap between the thought and the action, the itch will start to fade. So, feeding an addiction or compulsion just reinforces the addiction. The gap between thought and action is everything.

Monk: Yes, as we said when we were discussing thoughts, mindfulness is all about that gap, where we can learn to pause and make choices.

Ruby: I can't just let an itch go, I would go nuts. Sometimes, when I have an itch on my back, I have been known to rub myself against tree bark until I've sanded off all the bark, like a bear. Is that so wrong?

Monk: It's not the itch, it's the thoughts about the itch. The deepest addiction we have is to our thoughts. Of course, when you're withdrawing from heroin or alcohol, it's physical agony and you need treatment, but when you get through that phase you can start working on your mind. With mindfulness training, you can slowly learn to accept the itch and leave that gap before you go back to your habit of scratching. In that gap, you practise learning to accept the state of discomfort, not to push away the feeling. You're challenging the mind-set that tells you there will be relief if you scratch. Eventually, you'll be free of the itch.

Ruby: I still would need the tree bark.

Monk: Also, when I worked in the clinics, they talked about having a 'hole in the soul'. The addiction becomes a desperate attempt to fill some kind of deficit. But mindfulness practice is incredibly enriching and regenerative, and I think it helps a lot when they can learn to use compassion to nourish that 'hole' within them which led to the addiction in the first place.

Ruby: I think a nourished hole is everything.

You'll find the relevant mindfulness exercises for addiction in Chapter 11.

10

The Future

We keep talking about the future as if it's something out there that's coming. We're moving so fast it's already here. Your next breath is the future, so get ready. Google's chief economist, Hal Varian, says, 'A billion hours ago, modern *Homo sapiens* emerged . . . A billion seconds ago, the IBM personal computer was released. A billion Google searches ago . . . was this morning.'

New Evolution

In the past, we evolved primarily through our DNA. Each time the world threw a new challenge, yelling, 'Okay, sucker, how you gonna cope with this one?', our genetic makeup mutated to make sure we'd be able to see in the next New Year. Evolution has always come to the rescue. For example, about ten thousand years ago, to ensure the survival of Australian aboriginals living in desert climates, a genetic variant developed that meant man could now survive in boiling temperatures and take on a tan like nobody's business without burning. Another example is that, again about ten thousand years ago, when we morphed from being knuckle-dusters to standing-uppers, Europeans and Africans were the same dark colour: both came from the motherland – Africa. (A big 'sorry' to the fine racists of Alabama.) Over time, human skin in less sunny northern climes grew lighter, to help the people absorb the sun's ultraviolet rays and synthesize vitamin D more efficiently. Finding out that the colour they are is only because of the direction their forefathers walked in has bummed out a lot of bigots.

These days, we don't need to upgrade genetically because we've got things like central heating and sunblock. If we need to travel longer distances, we don't need to grow more or stronger legs, we just make faster planes or cars. Most of us don't go outside much, so here in the West one of the big challenges is our foot falling asleep from not moving it while you're sitting at your computer or getting a crick in our finger from texting.

Most of the natural selection today is because of cultural change, not climate change. Now, with technology at the tip of our fingers, and with the possibility of instant exposure to infinite environments and intellectual challenges, we just can't adapt our genes fast enough to keep up with life in the digital jungle. The reason we're burning out and sprouting new diseases is that we aren't equipped to be able to genetically modify that fast, and yet, if we don't, our species will go kaput.

To be able to keep up with the technological evolution, we'd have to evolve genetically a few times a month. Humans can produce a new generation only every twenty-five to thirty years, and it can take thousands of years to bake a new upgraded trait to be spread throughout the population. Some animals replicate once a week. There's a fly that lives for only twenty-four hours; at the end of a day of existence, they're replaced by a whole new generation. They're just getting into the swing of things and, *boom*, they're dead.

The New New Age

Probably as many science-fiction books will tell you, it was all part of the 'bigger plan' that we evolved to this moment in time to build computers that can take over from the old worn-out models called 'us'. What's coming in the future is coming. You can't stop evolution, and the technological add-ons that are here, or almost here, are becoming extensions of us.

One of the first marriages of humans and their technology will be repairing brain injuries or other cognitive dysfunctions. Good news for those of us with mental illness. Antidepressants, at present, splatter-gun the brain, randomly hitting any old receptor they come upon to alter your chemicals. Neural-stimulation techniques, coming soon to your future, will be able to focus on a single area.

Further down the line, there will probably be memory enhancement for the ageing. We already have deep-brain stimulation to alleviate symptoms of Parkinson's Disease, cochlear implants to restore hearing, and prosthetic limbs for those with disabilities.

Recently, electrodes to connect the motor cortex to the nervous system were inserted, enabling a quadriplegic woman to fly an F-35 fighter jet in simulation. A monkey used its mind to ride around in a wheelchair. (I'm not sure what the point of that one was, but he did it.) The scientist and physician Miguel Nicolelis and his team made it possible for a paralysed man to make the opening kick of the World Cup (see TEDTalks).

The same technology that allows a quadriplegic to use their thoughts as a remote control to move a bionic limb will make it possible for anyone to use their thoughts as a 'remote control' for everything. All your online shopping could be done simply by imagining it. The remote is already here and is being used by people who are paralysed, who can move a cursor on a screen just by using their thoughts.

As we speak, or as you read, work is being done by Elon Musk (owner of Tesla and SpaceX, who some say is brilliant, others not) and his team to create a brain–machine interface where all the neurons in your brain will be able to communicate with the outside world. Elon says, 'We already have a digital tertiary layer in a sense, in that you have your computer or your phone or your applications. You can ask a question via Google and get an answer instantly. You can access any book or any music. With a spreadsheet, you can do incredible calculations. You can video chat with someone in Timbuktu for free. This would've gotten you burnt for witchcraft in the old days.'

The New You

This century may be the one when we, as a species, manage to snatch the genetic code from the clutches of evolution and learn to reprogramme ourselves. People alive today could witness the moment when 'biotechnology' might be able to free the human lifespan from the will of nature and hand it over to the whim of each individual. So, now, a whole new set of questions come into play, such as, would you actually want to be immortal? And, if you do, where will the next generation live? Maybe in jars.

We've neutralized the power of natural selection with modern-day medicine and technological innovations. We don't have to wait around for new and more improved humans any more, thanks to in vitro fertilization. Parents can choose which embryo to implant, like choosing a lobster at a seafood restaurant. Specialists use gene-editing tools to create new mutations so parents will be able to design their own babies as far as gender, hair or eye colour is concerned. One of our big aims as a species is to develop greater intelligence, since it's what got us where we are today, and our genes have evolved to dedicate more and more resources to our brains. We won't have to wait for evolution to up the ante, we'll soon be able to enhance intelligence by choosing the most intelligent embryo. Also, we'll be able to manipulate DNA to engineer cells to create the next Einstein, Rembrandt or Olympic champion. I'm not saying this is good or bad, I'm just the messenger.

Kevin Kelly, publisher of the Whole Earth Review, executive editor at WIRED, founder of visionary non-profit organizations and writer on biology and 'cool tools', agrees that we might now start using the machines we've created to take the next step in our evolution. We are already working on implants for the deaf; the next step just might be that they can hear things which people with normal hearing can't, for example, the sound of a whale hundreds of miles away, or the ability to hear what someone might be thinking. Eventually, we won't be able to tell our software from our biological brains.

We Have Always Upgraded

Before we came up with language, a mere fifty thousand years ago, we had no way of getting a thought from your brain into my brain. Then the technology of language was invented, transforming vocal cords and ears into the first communication devices. We used these devices for years, and they seemed to work well, until we started to spread over large spaces and, then, no amount of screaming loudly would do the trick. Fortunately, phones came along to solve the problem, and everything since using our mouths and ears has been a technological upgrade, allowing us to stay connected to the world. I haven't heard anyone complain all these two hundred years since then, or send Alexander Graham Bell hate mail. Or troll him.

Let's face some reality at this point: twenty years ago, twenty thousand people had laser eye surgery to improve their vision; now, 2 million people a year get it done, it's no big deal and everyone's happy when they can suddenly read stop signs. Same thing happened with pacemakers and then organ transplants. There are waiting lists for organs, that's how widespread this technology is now. We have to accept that the brain–machine love match has already begun.

We've already got cyborgs roaming around (your next-door neighbour might be one). The definition of a cyborg in the dictionary is 'a fictional or hypothetical person whose physical abilities are extended beyond normal human limitations by mechanical elements built into the body'. As I said earlier, there are thousands of people walking around right now with cochlear implants, retinal implants, pacemakers, deep-brain implants, and so on. The number of robotic procedures performed increases by about 30 per cent a year.

I made friends with a guy called Neil Harbisson whom I met at TEDTalks Global and immediately dragged him to my home to meet the family. They thought he was the coolest person on earth. Neil was born with the inability to see colour, he saw only black and white, so he built an electronic eye that detects colour frequency and had it implanted in the back of his skull. Now, through bone conduction, he can hear colour through sound frequency. He hears the sound of red as C major. He says he can actually hear a Picasso or the sounds of a shopping mall and interpret these sounds into colours. Sometimes he puts certain colours of food on his plate so he can eat his favourite song. He has extended his senses by using technology as part of his body. One day, maybe we'll all be able to buy an implant and extend our own senses, just like your average superhero.

Guess What's Coming?

Ray Kurzweil is one of the world's leading inventors, thinkers and futurists, someone who talks about technology and trends and looks at the bigger picture. He says, 'Basically, thanks to the Human Genome Project, doctors are learning how to reprogramme the "outdated software" of our bodies.' He predicts, 'In the 2040s, humans will develop the means to instantly create new portions of ourselves, either biological or non-biological.' This means, if you feel like it, you can sprout wings – or have a bigger penis, depending on your mood.

These are all speculations, not facts, but rumour has it that, by the late 2020s, we'll be able to eat as much junk food as we want because we'll all have nanobots injected into our bodies that will provide us with all the proper nutrients we need while also eliminating all the excess fat we'll gain from eating twenty bags of Doritos and unlimited chocolate every day. Hurrah!

It's reported that, at some point in the future, we'll be able to beam ourselves into another person's brain and experience the world as they see it, just as in the film *Being John Malkovich*, or that one where these doctors shrink themselves down to nanometre size and go inside someone's body, riding through their bloodstream. They were almost eaten many times, by bacteria and foreign viruses. I saw it as a child and was convinced they were in me, so I sat on the loo for about a week, trying to squeeze them out.

It goes without saying, but I'm saying it: there'll be a

tsunami of digital sex. You'll be able to create an avatar who is blue-toothed to your genitals. *Voilà!* You won't even have to shave your legs or bother with makeup. You can be represented by Beyoncé while, in real life be a slob.

Enter the Robots

The biggest demand for human-like robots is coming out of Japan; everyone's living way too long and the older population isn't getting any younger. Today, 25 per cent of Japan's population is sixty-five or older and, by 2050, that's going to increase to 39 per cent; their Ministry of Health says that, by 2025, Japan will need 4 million caregivers. The birth rate in Japan is low and they don't like to let in too many immigrants, or give out work visas, so who's going to take care of Grandma/pa? Well, Japan is leading the world in the creation of robots so maybe they're looking for a solution there. Toyota has built one already, called Robina, named after Rosie, the cartoon robot in *The Jetsons*. Robina is just under four feet tall and can use words and gestures and wear a skirt. Her brother, Humanoid, can do the dishes, take care of Grandma/pa and even entertain, with specific talents. (I'm imagining the playing of spoons and clog dancing.)

Honda has created Asimo, a fully functioning humanoid, standing four foot tall with cameras for eyes. Asimo can answer questions, and, supposedly, interpret human emotions, movements and conversations.

My Conclusion

The drawback is that we might get so addicted to 'the next big thing' that we lose who we are and end up just being 'out there', communicating with virtual families and friends.

I hope we don't feel too lonely not being near real flesh-and-blood humans but I'm sure there will be some pill to take or some implant to implant if that happens.

A potential emotional sacrifice might be that if we 'have it all', we may lose out on any depth and just shallow out. Unless you experience a little bit of sadness or darkness, you won't be able to feel compassion for anyone else. This also means there will be no literature, because only the dark stuff can be great. No one ever won a Booker Prize for a peppy novel where everyone ends up at a picnic. On the other hand, in the future, there might be a button on your keyboard to delete pain and one to hit for compassion, so – problem solved. Let's also hope there's a button marked 'the present', otherwise you'll never be able to taste, smell, hear or see anything as it's happening live. It will all be recorded on video, where nothing is ever as good as the real thing.

Also, I hope that, in the near future, there might be an automatic button on the computer that takes you offline and starts making all the decisions for you, like deciding which events, parties and meetings you really need to go to, which friends are worth seeing and which are draining you, and tells you honestly what your 'look' should be, taking into consideration your age, weight and personality. This means you'll have time to have a life and declutter your brain. You can take it easy and it will do all the work. That's what I'm looking forward to. Otherwise, we'll continue to be slaves to the digital age.

One last thing, and this is the purpose and heart of mindfulness, I hope we'll still have the facility to pay attention, to focus on things we choose to focus on. Once you lose attention, you'll just get dragged from one thought to the next, which will scatter and rattle your mind. Unless we train ourselves to focus intentionally on what we choose to focus on – and, hopefully, it's things that make us feel good – we'll be in a constant state of distraction and dissatisfaction.

The Monk, the Neuroscientist and Me

Ruby: Ash, I was talking about Asimo, Honda's human-like robot. Will robots like that be able to recognize emotions?

Neuroscientist: Yes, in a way, computers can already be trained to distinguish between facial expressions. You show them lots of photos of people smiling and they learn what a smile looks like. Then the computer can categorize a smiling face as happy, and now you start to have a system that can recognize emotions.

Ruby: Does the robot know the smile means there's good news? Does it feel happy?

Neuroscientist: That's the trick. Emotional expression can be complicated and it's hard to cover all the possibilities with a robotic algorithm. With humans, we know that smiling might not always mean 'happy': there are genuine smiles, fake smiles and even angry smiles.

Ruby: How do you know a fake one when you see it? I'm smiling at you now, but can you tell underneath you're really getting on my nerves?

Neuroscientist: Yes, I can tell, because we both feel that way. We just pick up a vibe from each other. We're sensing many cues we're unaware of. When we have our own emotions, we become aware of our behaviour and that lets us recognize what other people are doing. Computers don't have that lived experience. They're not aware, they just detect.

Ruby: And I can imagine, if you're with a robot and it gets it wrong how you're feeling, that would really piss you off.

Neuroscientist: Yeah, exactly. My mother does that. She calls and tells me that I sound depressed but I wasn't before I started speaking to her.

Ruby: Ed can't read me either. He just stands there smiling and nodding but has no idea what I'm saying. I'm telling him I ran over his leaf-blower with the car and he's still smiling and nodding.

Neuroscientist: It's a hard problem, reading emotion, and it's something that humans get wrong all the time.

Ruby: I was reading that the Japanese are also introducing robots for childrearing, to help get more women in the workplace. It turns out that Japanese women are the best educated in the world but 70 per cent of them leave their jobs after their first child. The government and private funds are investing money in robots so the women can go to work. So, Ash, what are the chances that this chunk of metal can nurture a baby? What if we suddenly get robot babies? I wouldn't want to breastfeed the Terminator.

Neuroscientist: Would you breastfeed R2D2?

Ruby: Would you? 'Cos, in the future, that's what you'll be doing.

Neuroscientist: You know, scientists have been trying to make artificial mothers for a long time. In the fifties and sixties, an American psychologist named Harry Harlow did some amazing work on artificial mothers for monkeys.

Harlow's mother monkey was a block of wood covered with sponge and an old towel, with a light bulb on the inside to make it warm. He thought that all a baby monkey needed was food, water, warmth and something to cling to.

Ruby: I think I would have been better off if my mother was made of an old towel. She did, however, carry sponges wherever she went, in case a microbe of dust was visible to the naked eye. What happened to the monkeys with the sponge mothers?

Neuroscientist: Harlow thought it was a big success because, when the babies were frightened, they would cling to the sponge mother for comfort. But when we watch his old videos now, it's clear that the babies were so frightened they would cling to anything, even a large piece of lettuce.

Ruby: I know how they feel.

Monk: Yes, when I was brought up by wolves . . .

Ruby: Thubten's just upset because we've cut him out of the conversation.

Neuroscientist: (ignoring Thubten) So, clearly, these artificial parents aren't ideal. But it's a tough one for mothers because, exactly at the time it's most crucial to be there for the baby, it's also likely to be a critical time in their career.

Ruby: That's the bitch, that there's always this choice between giving up your job or giving up time with your kid. It's a lose–lose. What can the robot offer?

Neuroscientist: Robots can help with tasks around the house and maybe even with childrearing tasks like changing nappies, but the actual nurturing role would be very hard to mechanize. Caretakers mirror a baby's emotions, they teach babies how to self-soothe and calm down. Robots can't do that, not yet anyway.

Ruby: Okay, so no replacement for Mommy, but can they replicate a human brain yet?

Neuroscientist: I'm asked about this a lot – whether we can ever build a complete replica of a human brain. In principle, yes, the brain is a physical device that obeys physical rules, so we could build something that works like a brain. But the brain doesn't work in isolation. We're human because of our social connections, how we interact with each other. We see ourselves through other people's eyes, and that changes our behaviour. I don't think we can replicate that with technology.

Ruby: So, the computers would have to learn how to mingle? A cocktail party of computers spouting emojis at each other.

Neuroscientist: That sounds like my college reunion.

Ruby: Thubten, what are your thoughts about future technology? You're back in the conversation again.

Monk: Thank you! I work with some of the biggest technology companies. When I speak to them, I try to emphasize the need to look at solving the world's problems, not just making tech for the sake of tech. There needs to be a compassionate and ethical motivation. The driverless car, for example, won't be able to make ethical choices, it will just drive to survive. That's a concern, and I think the more we can get mindfulness and compassion training into those companies, the better chance there is of them creating a meaningful future for this planet.

Ruby: So, where does the human and the machine divide? I have a couple of capped teeth, and screws in my toes from a bunion operation. If I start replacing other parts on a big scale, when do I lose who the original me?

Neuroscientist: There is no original you. With the normal rate of turnover in bone cells, you create an entirely new skeleton about every ten years. Every single molecule in every cell of your body gets replaced many, many times over your lifespan. It doesn't matter if we replace our body with cells or with mechanical parts, that's not what makes us who we are. Do you know the paradox of the ship of Theseus?

Ruby: No, why would I?

Neuroscientist: The paradox was proposed by Plutarch in the first century. He describes the wooden ship that the Greek hero Theseus captained, kept as a memorial in the Athens harbour. Over the years, to preserve the ship, the Athenians replaced all the rotting wood plank by plank so that, eventually, not a single plank of the original ship was left. Was this still the ship of Theseus? Aristotle said it was, because the ship was defined by the idea of its identity more than by its materials.

Ruby: Ash. Stop showing off like you know Plutarch. I'll ask you an easier question. If you remove everything, like you take me apart, where is the essence of human consciousness?

Neuroscientist: That's the big question everyone wants to know, but I'm not sure I can even say what consciousness is, let alone where it is. When I work in the hospital, I'm just interested in conscious versus coma – and, by 'conscious', all I mean is whether someone can move and respond. I don't get into anything about self-awareness.

Monk: This is a central question in Buddhism: what is the self? Does it even exist? If it exists, where is it located? For something to exist, it has to have a location or other defining characteristics. So, in meditation, we explore the question, for example, if we were to remove a part of our body, does our self diminish by the corresponding amount?

Ruby: So, if I had full metallic body, where is the Ruby part?

Neuroscientist: We all know that losing a limb doesn't mean losing any sense of self. It turns out that you can lose quite a lot of the brain too, without losing any consciousness. For example, there are babies born with a large amount of 'water on the brain', or massive hydrocephalus, where the brain is crushed down to just a thin rim and the skull is filled mostly with spinal fluid. There is very little brain there but those babies can still have rich and complex behaviours.

Monk: The Tibetan philosophers talk about 'water brain', where the individual has almost nothing but water for a brain and yet is still conscious. So, this suggests to us that consciousness is beyond brain activity.

Ruby: So, how do we explain it?

Monk: We can't, because the mind can't understand itself through concepts. We can't be conscious of what we're not conscious of.

Ruby: Einstein said something like that. You stole that from Einstein.

Monk: Let's say we both said it. We don't believe you can pin consciousness down and, if we're constantly trying to define it, it's not getting us anywhere. Actually, the point of mindfulness is to directly see the illusory nature of our thoughts and the self. Breaking down the illusion of self will free us. My teacher often used to say, 'Don't take yourself so seriously,' which can be taken on quite a profound level. People experience this when they meditate. They start to identify less with their thoughts, and they discover an awareness which is beyond their pain, beyond their ideas of self.

Ruby: We were talking about the future, so can we steer the ship back to port?

Monk: In terms of the future, we keep talking about getting 'more advanced', but what do we mean? As a species, are we becoming happier, kinder, wiser? What we now need is to evolve our minds, upgrade the software. Maybe that's why mindfulness is so popular now.

Ruby: I wrote in the chapter on evolution that we evolve to deal with challenges the environment throws up. Technology is part of that evolution because we needed to deal with modern challenges. Now, the technology itself is the challenge, because it's making our lives more pressurized and it's got out of hand, so maybe our next step in evolution is incorporating something like mindfulness.

Monk: Yes, maybe mindfulness is the next step in our evolution.

Ruby: So, what we're saying is, we don't need any more thumbs but we do need a new kind of mind to overcome the damage we've done to the world. Do you think we can actually evolve our minds?

Neuroscientist: I hope so!

Monk: Because we seem to have made ourselves so unhappy, it motivates us to find a solution. In fact, the more messed up we are, the more we're motivated to meditate! Everyone wants a better life. There's a natural search for the solution.

Ruby: We should build better minds, not better missiles. I've become a hippie. Let's light up a joint and celebrate.

You'll find the relevant mindfulness exercises for dealing with the future in Chapter 11.

11

Mindfulness Exercises

The Benefits of Mindfulness

All the topics in the prior chapters lead to the practice of mindfulness. Practising it is the only way I know to be able to find some peace in a world that's not peaceful.

Mindfulness isn't a spa treatment, like bathing in a warm, sacred urn of Nepalese yak oil, it's hard-core – Iron Man for the brain. It takes stamina and commitment to build up those brain muscles to make them strong enough to rope in that wild, woolly mind; otherwise, it will run you ragged.

Here are some of the benefits:

1. **Breaking bad habits**

   Personally, mindfulness has helped me curb my addictions, one of which is anger. Rage is my drug of choice. With mindfulness, I've recognized it's a bad habit and that all I get from indulging my fix is acid reflux. We all have bad habits of thinking and feeling, and mindfulness is a way of recognizing them, forgiving ourselves for them and nipping them in the bud.

2. **Stress reduction**

   I can't mention cortisol enough. Like it or not, empirical evidence shows that mindfulness reduces anxiety, panic and stress. You can't argue with science.

3. Longevity

I don't think I'm being unreasonable when I say I'd like to live a long, healthy life. Overpriced moisturizers and facial reconstruction don't cut it because, obviously, your insides still age and there's far more of you inside than outside. There aren't enough surgeons in the world to nip and tuck it all.

What defines your real age is the wear and tear of your telomeres. How long they are determines how fast you age, so, if you lie about it, your telomeres will expose you. The scientists who discovered them won a Nobel Prize. Without getting too 'sciencey' on the subject, telomeres reside at the end of your chromosomes, which you have in every cell that makes you *you*. They're like the little plastic bits at the end of a shoelace that stop it from fraying. Whenever your cells replicate as you age, telomeres get frayed and wear down. If they become too short, they wither and die. How fast they fray depends how you live your life. Research shows that mindfulness helps you live longer and stay healthier.

I haven't had my telomeres tested but I feel, in my bones, a lot younger than I really am. To me, an inflexible mind and body is a sign of ageing. At this late stage in life, I'm proud to say there's almost nothing I won't try; I've been to the 'Burning Man festival' three times and I can still do the splits. Case closed.

4. More visits to the present

Most of us spend about 50 per cent of our lives mind-wandering; sometimes we have nice thoughts but, mostly, they're negative: rehashing and worrying about things that have or haven't happened. I figure I've missed enough of my life; I don't want to miss any more. I practise mindfulness so I can have a front-row seat to watch my life with no intervals.

You can take as many selfies as you want of yourself in front of a chocolate brownie, but nothing compares to that firecracker going off inside, blasting out pure pleasure dust, when you've got that brownie in your mouth. We live for the moment, but no one tells us how to get there. Mindfulness trains you to stop and smell the roses.

5. **Better attention and memory**
   Now that I practise mindfulness, I have better control of flipping my attention to where I want it and away from where I don't, even in the face of stress. When you're under pressure, your memory goes down and you go blank. I'm happy to report that, during the tours of my one-woman show, in over two hundred performances, I've never forgotten a line.

Thubten and I are now going to give specific mindfulness exercises to help you deal better with your thoughts, your emotions, your body, with compassion, in relationships, with your kids, with addiction, with the future, and with forgiveness. There are many different ways to teach mindfulness, so Thubten and I are offering you a selection. As in everything, different styles suit different people.

First of all, some general points from Thubten which will be helpful when you do any of the exercises that follow.

1. **Find a quiet place** for your practice.
2. **Practising first thing in** the morning is ideal, as it means you're starting your day off right, but really, any time you can fit it in is great. Also, it's good to practise

micro-moments of mindfulness a few times a day in addition to your regular 'formal' session.

3. **For beginners, the length** of the session should be five or ten minutes. Later in your practice you can choose to up the timing. Use a clock or a timer. It's not the length of the session that counts; practising every day is what changes the mind.

4. **It's good to start** and end each session with a moment to set the intention of kindness. Remind yourself that you're practising to help yourself *and* others; as your mindfulness grows, the effect you have on others will become more and more positive. This is compassion.

Mindfulness Exercises for Thoughts

THUBTEN'S EXERCISES

During mindfulness practice, the mind will wander a lot, but this isn't a bad thing, it's a chance to exercise. When you get distracted by thoughts, you can bring your attention to the body or to your breath, coming back to them as an anchor when your thoughts have carried you away. Later, as your practice develops, you might not need to anchor using either the body or the breath, you'll be able to just stand back and let the thoughts pass by. They won't disappear, but they won't be so intrusive and overwhelming.

It's important to understand that you're not trying to get rid of your thoughts, you aren't trying to go blank – which is impossible, anyway. Instead, you're focusing your mind so that you can be less controlled by your thoughts. The point of the training is to enable you to return again and again to the present moment.

It's difficult to steady the mind, so when you notice you've been pulled into mind-wandering – the past, the future, your fantasies, distractions such as external noise, and so on – don't be self-critical. Mindfulness is all about learning to be kind to yourself. So, when you notice that your mind has become distracted, just gently come back to the area of focus.

Exercise 1: Body Scan

- Lie on your back on the floor. If it's more comfortable, place a pillow under your head and knees.
- The next step is to get a general sense of your body lying still. Feel the floor beneath you.
- You'll be moving your focus through specific areas of the body to hone your attention. Try to sense the sensations directly, rather than think about them. If your thoughts are scattered and your mind is distracted, don't get frustrated, be kind to yourself and bring the focus back to the specific area you are working on.
- If you can't sense anything in an area, just be aware of the lack of feeling.
- Start by focusing on your toes (left and right feet simultaneously).
- Let go of that focus and move up into the soles of both feet, again, not thinking about it, just sensing the whole area, from the tips of the toes all the way to the heels.
- Let go of that and move your attention to the ankles, feeling that entire area.
- Slowly shift your attention up through the calves and towards the knees. Really sense the whole knee: bones, muscles, skin, etc.
- Now move to both thighs, being present and sensing that region.

- Next, bring your attention to the pelvic area and then slowly up to the waist.
- Now let go of that and move your focus, like the beam of a spotlight, into the fingers of both hands.
- Slowly shift your focus to the hands, then up into your arms, elbows, armpits and shoulders. People often feel tension in their shoulders and, once more, it's important not to be critical, just focus on whatever's there.
- After the shoulders, move your focus to your torso, travelling step by step, upwards from the waist, then the lower back, up to the rib cage, chest and upper back.
- Now, moving up to the neck, notice whatever's there. If you feel any tension or stiffness, don't judge it.
- Next shift your focus to specific areas of the face: the mouth, cheeks, nose, eyes, eyebrows and forehead.
- After this, move your attention to the top of your head. Spend a few moments resting your awareness there.
- The next phase is to reverse the flow of attention from head to toes, but this time more quickly and in less detail. Use a sweeping motion, more like a water line descending, as if you were pulling the plug out of a sink. Travel quite swiftly down from the head to the shoulders, torso and arms, to the legs, feet and toes, noticing each area.
- When you've finished, spend a few moments relaxing, having a general sense of your body lying on the floor, feeling it beneath you, and noticing your breathing, without manipulating it in any way.
- Some people may find that this body scan makes them feel dizzy or disorientated. If this is the case, do the scan only from head to toes, first slowly, and the second time a little faster.

Exercise 2: Breathing

Sit on a chair with your back straight. Try not to lean against the back of the chair; if necessary, put a cushion behind the base of your spine for lower-back support. Make sure your head is upright and your face, jaw and shoulders are relaxed.

This exercise has four steps.

Step 1. Body awareness. Feel the contact between your feet and the floor. Then move your attention to the sensations in your lower spine and bottom, where they make contact with the chair. Next, bring your focus to your shoulders. Now move into your hands and fingers and feel them resting on your lap. This brings your attention into the present moment, making you feel relaxed and calm.

Step 2. Notice your breathing, initially just the feeling of your body moving slightly with each in breath and out breath. Be careful not to manipulate your breathing; leave it natural, as it is.

Step 3. This is the most important step, so spend the most time on this. Narrow your focus to the feeling of the breath coming in and out of the end of your nose. Feel the air brush against the skin at the edge of your nostrils, again being careful not to manipulate your breathing. If it's uncomfortable to breathe through the nose, then breathe through the mouth and focus on the air as it crosses your lower lip. The nostrils or lip is the point of focus that you will return to whenever your mind wanders. Your mind *will* wander. This happens to everyone and is normal, but just keep bouncing the awareness back to the breath. Don't be frustrated by distractions; all you need to do is notice when your thoughts have taken you away and then gently return to the breath with patience and kindness. Sometimes, your mind will wander for a long time before you even notice that it has gone; that's okay, your focus will improve with training. It's the noticing and the gentleness of returning that count.

Step 4. When you're ready to end the session, go back to focusing on the feeling of your body where it contacts the chair, and the sensation of your feet on the floor.

RUBY'S EXERCISES

Exercise 1: Three-minute Thought Exercise

There are three steps in this exercise.

Step 1. (One minute) At any time in the day, or right now, notice what you're thinking about. If you're thinking of something good, bad or neutral, notice it, but try to do it with curiosity, not judgement. Even if your mind berates you for always focusing on the negative, at least congratulate yourself for noticing – most people haven't got a clue when their mind has wandered.

Step 2. (One minute) Take your focus to the breath, zooming in at the tip of the nose, the throat, the chest or the abdomen.

Step 3. (One minute) Widen your focus to your whole body, breathing in and out; imagine you're a bellows and the air is filling your whole body on the in breath and emptying it out on the out breath. You might notice your thoughts have receded slightly (they never disappear, the thoughts are just less intrusive). They become more like background noise, as if you're listening to a radio that's on in another room. When they're not so overbearing, you have a space to notice them and not be so drawn in by them. Eventually, you'll no longer be at their beck and call. You'll be able to sit back and watch.

Exercise 2: Sound

Just as in Thubten's exercise on breathing, sit on a chair with your back straight, but not rigid. Feel where your feet and bottom make contact with the ground and the chair. Let the focus on those sensations go and bring your focus to whatever sounds you hear. Don't try to find them, let them come to you; then listen to the pitch, volume, quality of the sounds and the silence between them. When your thoughts snare you, don't feel you've done anything wrong, just bring your focus back to the sounds without beating yourself up that you've done something wrong. Listen to the sounds; don't analyse or label them. Simply observe them for a while and then bring your focus back to the contact between your feet, the floor and your bottom on the chair.

Exercise 3: Bubbles

Do this whenever you feel your thoughts are going on a rampage, when you can't think clearly or you're caught in a brain fog. Imagine your thoughts as if they're thought bubbles, like the ones in cartoons. Each time a thought comes up, imagine it being inside a bubble outside your head and picture reaching out your hand and gently popping the bubble. It's so easy, I figure, why don't I live my life doing this? You just go, 'Pop, pop, pop.' And the thought bursts and vanishes.

Exercise 4: Refocusing When Your Mind Leaves Town

(This is almost impossible to do, but I find it amusing.)

- You can do this exercise while you're reading these words. Keep reading the words in this book . . . that's it, you're doing really well. Are you following what I'm writing, or has your mind gone into your own inner story and you're thinking about something else? If not, keep reading, but if you notice your mind has gone elsewhere or these words aren't making sense, stop trying to read and take a second to notice where your mind went. Did you go blank? Did you get carried away in a thought storm or think about yesterday or what's happening tomorrow? If you're at home, did you forget the whole exercise and find yourself in the fridge, looking for a chicken leg? Once you've noticed that your mind has strayed, bring your focus back to this page and start reading again and, to reward yourself for noticing that your mind went on a short holiday, get up and get yourself another chicken leg.

THUBTEN'S EXERCISES

The body scan and breathing exercises for thoughts also work for emotions. When an emotion pulls you away, just like a thought, all you need to do is bring the focus back to the physical body or your breathing, without judgement.

Exercise 1: The Sky

- Sit somewhere outdoors where you can look at the sky, or indoors near a window where you can see the sky.
- Your eyes are open, looking at the sky. Breathe in and out deeply three times. Imagine you're breathing out all your painful emotions into the sky and they are dissolving into that open space. You might feel lighter and unburdened. After these three deep breaths, return to normal breathing.
- Continue to look at the sky. Don't try to focus on the breath, just be aware of the sky. Notice whether you're gazing blankly, your mind is thinking of something else or you're getting lost in thoughts or emotions. When that happens, gently bring your focus back to the sky. Your mind is focused, in a relaxed and open way. As you're looking into the vast sky, see if you can experience that same expanse inside yourself. Simply look at the sky and let your emotions melt into the expanse.
- If there are clouds in the sky, feel you're looking beyond them. Just as the clouds aren't solid, neither are your thoughts or emotions. You don't need to become involved in them, you can simply let them go. Your mind is bigger than your emotions, just like the sky is bigger than the clouds. The sky is unaffected by the clouds, and your

mind can be unaffected by emotions. If your eyes start to get sore, close them from time to time.

- At the end of the exercise, bring your attention to your body (for example, focus on where your body is in contact with the seat) for a few moments, to become grounded again.
- This practice is especially good to do when sitting on a beach, noticing the expanse while looking at the horizon where sea and sky meet. If there are waves, let them come and go, just like your emotions can come and go without you needing to hold on to them or get involved. They are just emotions, nothing solid. The ocean is bigger than its waves, just like the mind is bigger than its emotions. When you can relax into an open state of awareness, you don't need to feel bothered by your emotions.

Exercise 2: Focusing on Feelings

- Sit somewhere quiet, in a comfortable position, back straight but not rigid, head and shoulders relaxed.
- Start to notice your breath, where you sense it moving in and out. Don't try to control it, but allow it to go at its own pace. After a few breaths, scan your body for any emotional feeling. Maybe there's a feeling of sadness, upset, worry or fear. See if you can take your focus to the exact location of the sensation, feeling its shape, edges, depth, weight, and so on. If you start thinking about the emotion, or the storyline behind it, gently bring your focus back to the area. As you pay attention, notice if anything changes. Maybe it starts to dissipate or change location. This helps you realize that emotions aren't solid.
- Try to drop the storyline that comes with the emotion: 'He said this' or 'She did that', and so on. If the mind flies off

into those stories, keep bringing it back to the emotional feeling in the body. You're relaxing into the feeling, befriending it, and as you do, the resistance starts to drop away. This exercise is training you to discover a non-judgemental acceptance and, ultimately, freedom.

RUBY'S EXERCISES

The idea behind mindfulness isn't to learn how to turn off emotions but to learn to be okay with them, no matter how hard the punch.

Exercise 1: Tagging the Feelings

- You can do this on the bus, in a cab, on the Tube, while bicycling, in the shower, or really anywhere.
- Follow the feeling of your breath going in and out. If you sense an emotion, try to hone in on the area of your body where you feel it most. Try to label the feeling with whatever word comes into your head. Choose *one* word that feels most relevant; don't make up a whole story. When you label an emotion, a space opens up around the feeling and the intensity of the emotion lowers. You're stepping back from the feeling and becoming an observer of it. Some mindfulness teachers say, 'Name it to tame it.' You can repeat the word as long as the sensation stays and, if the emotion changes, give it a new label.
- Remember that you aren't trying to make the sensation go away, you're just sitting with it like you would with a friend who's struggling and you're helping them find the right word that says it all. And if the emotion becomes too intense, just go back to bicycling or the shower or go eat another chicken leg.

My Story

A few months ago, I woke up gripped by the throat with anxiety. In the past, I wouldn't have been aware of what the feeling was and, for the rest of the day, I'd try to hunt someone or something down to pin my anxiety on and blame them for it. This time when I woke up, I took note of my inner state and gave it the label 'anxiety'. This was my internal weather condition; it hadn't been caused by anyone or anything. After doing some mindfulness practice, I realized I was anxious because of the dream I'd had the night before: A moose was chasing me down Kensington High Street. Once the feeling rose and I could label it 'anxiety', the emotion receded. It was such a relief. I could even laugh about it. I thought, *Ha, ha! Why would a moose be on Kensington High Street? Isn't that ridiculous?* (If anyone reading this has been chased by a moose down Kensington High Street, you have my sympathy.)

Exercise 2: Change Your Posture, Change Your Brain

Your bodily state is a reflection of your emotions and thoughts, and vice versa. When you let go of tension in the muscles, the emotions and thoughts also loosen up. If your body is tense, so are the thoughts and emotions.

Step 1. Hunch your shoulders, look down, frown and walk slowly, even shuffle. Shorten your breathing. Notice the influence it has on your feelings and thoughts. Don't do it for too long, or you'll get upset and blame me.

Step 2. Stand up straight, chest out, walk confidently and smile. (No one's looking, just do it.)

Feel the difference? You can't snap out of chronic depression by changing your posture but, if you're feeling a little blue or anxious, play with your posture by deliberately changing it.

You'll notice that, when you change your body posture, people who see you sometimes change theirs. We all catch each other's moods through signals in the body. So, spread your happy body!

Note: When You Hit Depression

I've always believed that, when I slip into the depths of depression, it would be cruel to try and do any kind of therapy, let alone mindfulness. I use mindfulness to be able to sense an oncoming depression but, if the floodgates open and it hits hard, then I know to back off.

People are always making suggestions when you're depressed, saying things like, 'Have you tried perking up?' The words 'fuck off' come to mind. How can you possibly tune into your mind when your mind is gone? If you could tune in, you'd most likely hear the voices of hell, because that is a symptom of the disease. In my opinion, all you can do when you're chronically depressed is wait it out and, if you're lucky enough to find the right medication and lucky enough to have compassionate friends and family, you've scored.

Once the light starts to shine, even if it's only a narrow shaft, that's when you could try practising a little mindfulness. Focus on the body or the breath, allowing the thoughts to pass by, remembering, above all, to be kind to yourself as you bring the focus back. Don't do this for more than five minutes at the start. And if you begin to get overwhelmed by the thoughts, do not continue until you're feeling better. As you come out of the depression, or whatever mental illness you may suffer from, go back to your usual practice.

Mindfulness Exercises for the Body

THUBTEN'S EXERCISES

Using our body for mindfulness helps to bring our practice into our daily life, rather than just leaving it at home on the chair.

Exercise 1: Three Steps, Ten Days Each

This is a programme you can practise over a month and is a good way to build up mindfulness in your daily life.

Step 1. For the first ten days, choose two or three simple activities that you do every day, like washing your hands, brushing your teeth, walking, and so on. Decide that these will be your 'mindfulness activities', you'll be mindful while you do them. This is a great exercise, as it gives you something physical to hook on to and will instil the habit of being mindful in general.

Every morning when you wake up, remind yourself of those two or three activities and your commitment to do them mindfully. (Stick with the same activities throughout the ten days.)

Maybe one of your chosen actions is brushing your teeth: we usually do this in a distracted way, but now try to focus on it with full attention. Sense the bristles against your teeth; taste and experience the toothpaste and water. Try not to run off into mental commentary. If that happens, bring your awareness back to the sensations. It's a mindful moment; you're staying present and focused.

Step 2. For the next ten days, rather than focusing on specific actions, practise 'micro-moments' of mindfulness for a few seconds many times each day, like little drop-ins. Wherever you are, at home, at work, travelling, walking the dog, and so on, notice the sensations in your body – check in, without judging anything.

Notice the sensations in your shoulders, back, face, feet or any other parts of the body, even if you're in a busy situation. You don't need to freeze or stop doing what you're doing but, just for a moment, for example, take your focus to your shoulders (maybe that's where you hold a lot of tension). Just by noticing the sensations without judgement, you start to relax.

Step 3: For the concluding ten days, continue practising mindful moments throughout the day but now add another layer: be mindful whenever you're waiting for something.

What do we often do when we're waiting? We check our phone obsessively, or gulp food down while texting with the other hand, or our mind has gone to Hawaii. Waiting offers a very powerful opportunity for mindfulness training, and it's useful because, in general, when we're waiting for something, we become tense or impatient. People get frustrated when standing in a queue, being stuck in traffic, or when the internet connection is slow. So, while you're waiting, let go of any stress or feelings of impatience by bringing your focus to your body. Feel the ground beneath your feet (I do this a lot when I'm on the Underground in London during rush hour), or feel the chair under you, or maybe sensations in your shoulders, stomach, that you've become aware of. When your mind flies off into frustration or other thoughts, bring it back to the body. Again, remember not to judge the sensations, just be present.

This is very good training for being able to deal with the more challenging things life might throw at us. We're starting with easy things like waiting, and this will help us stay steady in the bigger, more stressful moments in life. We're reprogramming our habitual reactions and in doing so we're rewiring the neural pathways in the brain. Normally, we would have an automatic reaction, falling back into old habits. Mindfulness teaches us a different way of responding.

A traffic jam has become like going to the gym. You can think, *Bring it on, I can use this to rewire my neurons.* It's all a training opportunity, teaching you how to be happy against the odds.

Exercise 2: Pain

- The million-dollar question for many people is, how can they practise mindfulness when they're in physical pain? They're afraid that, if they focus on it, the pain will get stronger, but pain isn't just physical, it's also the mental reaction of pushing it away, which creates more tension: pain about the pain.
- In this exercise, focus in on where in your body you're feeling pain. Try not to get caught up in your thoughts about it, just feel it directly; use the pain as the object of your mindfulness. When the mind runs off into thoughts, gently come back to the direct focus on the raw sensation of the pain. If at any time it becomes overwhelming, just focus on your breathing for a while as in the earlier exercises.
- Using pain as a mindfulness object helps us stop feeling tense about the pain as we learn to consciously focus with less judgement.
- The pain might not go away, but your relationship with it changes because you're no longer adding the mental pain on top of the physical pain. The physical sensations might start to move, become less solid, or you might start to relax around the pain, not being overwhelmed by it.

Exercise 3: Fatigue

- Mindfulness can also be used to help you cope with tiredness or severe fatigue. As with pain, we can learn to stop pushing these sensations away, stop feeling stressed about them.
- When we're tired, we are in fact also quite tense, because we're holding ourselves up – which, in turn, makes us more tired. In addition, we may have a lot of self-condemning thoughts, for example, *I'm so weak. I really can't face this*, and once that happens, the tiredness is no longer just physical, it's become a mental battle too.
- In this exercise, locate where in your body you feel the fatigue (or perhaps it's in your entire body), and let your awareness settle into that area and that feeling; move closer to it rather than pushing it away. Focus on the sensation of fatigue, using it as your mindfulness object, and when your mind drifts off into thoughts and emotions, gently bring it back to the sensation in your body.
- Sit in a chair, or lie on the floor or in bed. Feel the support of the chair, floor or bed beneath you – they're doing the holding up, not you. Usually, when we're fatigued, we feel as if we're having to hold our body up with effort and that's exactly what the feeling of tiredness is. Instead, we can relax and let go into the chair or bed which supports us. Go through your body and feel little micro-sensations of tension and let them drop into the chair or bed.

RUBY'S EXERCISES

Exercise 1: Walk the Talk

- You can use mindfulness with any physical exercise. It's always about noticing when your thoughts or emotions pull you away into distraction and then using the movements in your body as a point of focus. When you're walking, for example, try to sense what each individual movement feels like, contractions, stretches, strains, releases in your muscles. With body awareness, you automatically reduce the possibility of rumination. Practice gives your brain a break.

- Do this exercise for a few minutes at a time, and make sure you don't walk into a lamp post.
- Drop your attention down to the soles of your feet where they contact the ground. Focus on the sensation. Then notice the feeling of lifting your right foot up, the swing of your hips and exactly which part of the right foot touches the ground first and last. (Heel to toe is the way most people walk, though ballet dancers walk toe to heel, but what do they know?) Now do this with your left foot, lifting it, swinging it forward and being aware of where each area of the foot touches the ground. When your mind snatches you away, bring your focus to the last point of contact of your foot with the ground.
- In time, you can do this more quickly. You can also use this exercise when you run, skip, dance and hula hoop.

Exercise 2: Urge Awareness

- Do this when you notice yourself starting to rush; the urge to speed up. The more tuned in you are, the more you'll be able to sense if you're pushing yourself too hard. If you focus exactly where you imagine you feel the urge in your body, the mental urge disperses. Also, this exercise isn't to teach you how to move more slowly but to sharpen your attention muscles. Any time you focus on the body you enter the 'now', and this leads to a clear and open mind. This is also when you're at your most creative. Who knew that using your body could make you more creative? I did.

- Notice your movements when you're going into a difficult meeting or somewhere you don't want to go and compare them to how you feel when you are going to a party or meeting a friend.
- When you feel the impulse to speed up, try to pause intentionally and notice what's going on in your mind. Do this with curiosity not condemnation, thinking, *Oh, no, I'm racing again, I'm an idiot.* Some people find this exercise agonizing to start with, because your mind always wants you to speed up. So, when you notice it revving up, pause, breathe and proceed. Eventually, you won't need to pause consciously; it will become a new habit.

THUBTEN'S EXERCISES

Exercise 1: Compassionate Body Scan

- You can do this exercise in a quiet place, either sitting in a comfortable chair or lying on your back on the floor. If you're on the floor and you have a bad back, put pillows under your knees and head. Feel supported by the ground or chair beneath you and relax completely, feeling held.
- Spend a moment setting your intention in this exercise, generating the motivation of compassion. Only by working on compassion for ourselves can we then develop compassion for others. Remind yourself that you're doing this practice to build acceptance and kindness. Often, we judge ourselves, thinking our bodies should be thinner, younger or more beautiful, but with this compassionate body scan, we're learning to accept and be kind to ourselves. The acceptance leads to an openness; there's no longer any pushing for results.
- You're going to slowly scan from the head down to the toes, spreading a feeling of compassion throughout your body. Do this by imagining a warm liquid balm slowly filling you up from top to toes, starting at the crown of your head and moving downwards. Imagine either liquid or white light filling you with self-acceptance and compassion, soothing any feelings of anxiety or stress.
- If at any point you feel a difficult emotion or physical pain, just be with those sensations in the same way you'd sit with a friend who's in pain and you're there for them. If you feel no particular sensation, just be with that.

- Starting at the crown of your head, visualize or imagine liquid or light there, spreading downwards.
- Next move your focus to the eyes, then the face muscles and, finally, the mouth. If you're holding any tension in the face, let the liquid or light go into those areas and relax them.
- Bring your attention to the jaw, then the shoulders and armpits. Many people feel tension in their shoulders so, when the liquid or light reaches that region, let it soothe any feeling of discomfort.
- Move down the arms and body together, until you reach your abdomen and lower back. If you notice any physical or emotional tension in the abdomen, again, feel the balm moving through, generating a sense of acceptance.
- Move down the waist, into the buttocks and pelvis, then travel down the legs (both together), and into the ankles and feet. End with the tips of the toes.
- Finish the exercise by again simply feeling the floor or chair holding you. Your body by now is completely filled with the soothing balm or light. Then let go of that image and notice your breathing, keeping it natural.
- Finish the exercise by generating the wish to be of benefit to yourself and to others. This is a moment of compassionate intention.

Exercise 2: Breathing for Compassion

- We can begin by working on compassion for ourselves and then develop compassion for others.
- Sit upright in a quiet place with a straight back. Set the intention that you're practising compassion.
- Spend a few moments being aware of your body: feel the ground beneath your feet and the contact between your body and the chair. Be aware of your shoulders and let them relax.

- Focus on your breathing without trying to control it. As you breathe in, imagine that your breath is going directly to any area in your body where you feel discomfort. Maybe it's a physical problem, or maybe you can feel a sense in your body of where you're emotionally troubled.
- Imagine the in breath is bringing light into that area, soothing it.
- As you breathe out, imagine that dark clouds of smoke are coming out of that area, exiting the pores of your skin and dissolving into the space around you, freeing you. Keep repeating this cycle. Remember to leave your breath natural, don't force it. End the session by relaxing, feeling the chair under your body and the ground under your feet.
- Finish with a moment of compassion, generating a sense of kindness.

Exercise 3: Breathing for Compassion for Others

Use the former exercise for yourself, but if you're with someone or imagining others who are suffering, you can send compassion to them.

- Do the previous exercise, and then after you've inhaled soothing light into your area of pain, imagine sending the light to the other person(s) and filling them up with it, as you breathe out.
- Keep gently repeating this cycle.
- This exercise transforms our instinctive tendency to resist or avoid emotional pain; it means you're breathing in compassion for yourself as well as breathing out compassion for others.

- Eventually, begin to imagine more individuals, so that you are extending the compassion on a greater scale. You can even include individuals you normally feel aversion to. This helps the compassion become more and more unconditional.

RUBY'S EXERCISES

Exercise 1: Swinging from Negative to Positive Made Easy

- At any time in the day, or right now, notice what you're thinking about. This moment of self-reflection will give you the greatest insight into what your habits are. If you're always thinking about something that makes you feel good, you can skip this exercise, you're doing just fine. If you notice that you tend to focus on things that make you feel sad, bad or blue, stay with your thoughts but explore them with curiosity, not judgement. Even if your mind berates you for always focusing on the negative, at least congratulate yourself for paying attention; most people haven't got a clue. (Remember: we all tend to focus more on the negative, so forgive yourself. It's a habit from the past, trying to help you stay in the gene pool.)
- Now, bring in a thought of something or someone (it doesn't matter how small the detail) that makes you feel good – lighter, perkier. Even in the darkest thoughts, there can be a ray of light. (In my case, it's picturing my cat, Sox.) Notice the feelings in your body; gradually you'll become more conscious that, when you change your thinking, your inner landscape changes.
- Note: Please don't think this is about putting on a big smile and skipping in the daisies when you feel down, it will only make you hate yourself more for failing at

it. We have to accept the negative because it lives within us, but if you can bring up those positive memories even for a few seconds each day, the brain starts to rewire and the habits of continuous negativity begin to break down.

Exercise 2: Using Sounds to Make You Feel Good

I know this one seems weird but, sometimes, when I'm caught on a real stinger of self-hatred/abuse/criticism, I find some earphones and listen to recordings of the sounds of tropical rainforests (see Spotify). As soon as I hear the squawk of the macaws and pitter-patter of rain on the leaves, I either get serene or I have to go to the bathroom.

Exercise 3: Kindness: The App

I'm going to pass this one over to my son, Max, who's made a free app to help people develop habits of kindness using tech to make the world a better place.

MAX

'I created an app that makes performing regular acts of kindness as quick, simple and fun as possible. This is done with weirdly wonderful cartoon characters who celebrate your progress with tips to keep you going. There's also a challenge mode and a feature that introduces you to the flow of compassion, led by Professor Paul Gilbert. If you're stuck for ideas, you'll find over one hundred here, and you can swipe through these suggestions – sort of like Tinder but, instead of finding singles in your area, you're making the world a better place. When you perform kindnesses, there are surprise rewards, though nothing is as rewarding as doing the act itself.

In recent years, kindness has got a bad name. I googled

"kindness" and up came a TEDTalk given by an Orange County housewife who remembered the time she gave a homeless man a hot dog and they cried in each other's arms, until she had to go to her next pedicure appointment. I've worked hard to keep the schmaltz to a minimum and make the app as practical and simple as possible. Also, if you wake up feeling nasty, there's an option to let rip and just be a badass.

This app isn't a business opportunity, it's just a way I think technology can help people. It's currently available and free to download and use: search for "The Kindness App" or go to thekindnessapp.com. I would also like to thank my mum for her act of kindness in donating me this space to tell you about my app!'

Exercise 4: What's in a Face?

Professor Tania Singer conducted research into the neuroscience of compassion. In a clinical setting, over several weeks, she asked the volunteers to look at a computer screen showing expressions on people's faces. The computer collected the data by people tapping on the keys when they registered angry, fearful, anxious or joyful faces.

Throughout the weeks, they were encouraged to focus on the happy faces until this became their new default. Their brains were scanned throughout the experiment and it became evident that different areas of the brain were activated when going from a negative tendency to a positive one. When they started to pay more attention to the happy faces, chemicals were produced which boosted the immune system. Blood and glucose were increased in the brain, energizing it and promoting a sense of ease and well-being.

Obviously, you haven't got a brain scanner at home, so here's how you can change your default from negative to positive.

- When you're walking down the street, at the office, at school, on a bus or anywhere in public, notice who's smiling and who looks upset. Over time, intentionally send your focus to people who seem light and content, allowing their emotions to rub off on you then sending it back. (You don't have to talk to them just take it in and then send back the vibe.)
- For those people who look unhappy, stressed or anxious, try to imagine what's going on in their lives. Notice their faces, posture, where they're holding on and try to tune in to how they might be feeling. You don't have to get it right, it's just for you to exercise those empathy muscles. The intention is everything.
- The reward of these exercises is that when you're actually with a friend or someone who's having a hard time, you're tuned up to listen with empathy. Once you can clearly read their state, you'll be more accurate in how you can help them with compassion. Not just saying, 'Get well soon.'

Mindfulness Exercises for Relationships

THUBTEN'S EXERCISES

These are all exercises aimed at creating a shift in our attitude or way of thinking, which is an important part of mindfulness training.

Exercise 1: Writing a Letter

- For this exercise, sit somewhere quiet and meditate on your breathing for a few moments, just to focus the mind and bring yourself into the present moment.
- Then write a letter (don't send it!) to the person you're experiencing a difficult relationship with. In this letter pour out exactly how you feel. Don't hold back.
- After that, write a letter as if from them to you, not thinking that they've read your letter, but just let them pour out all their feelings.
- Next, spend a little time looking at both letters, trying to generate understanding and even compassion for both parties.
- The point of this exercise is to gain perspective through exploring both sides of the story. You're learning to be a neutral observer, not getting hooked into storylines.
- Once you've read both letters, sit and meditate again for a few moments; don't focus on the breathing, but let any feelings arise freely in your mind. Don't push them away, just experience them.
- Lastly, write another letter from you to the other person, in which you constructively express all the ways you'd like to move the relationship to a more harmonious place.

Exercise 2: Dyad Exercise

- This is a dynamic practice involving two people. It can be done face to face or by telephone. A dyad is an intentional and structured meeting with the aim of developing empathy and compassion. Each person is going to speak for five minutes without interruption.
- The first person talks about something difficult that happened that day, and how it felt. They should list all the ways it made them feel, in their mind and body. Then they talk about how they felt about something which happened that day or recently for which they feel grateful.
- The listener simply maintains eye contact (unless the exercise is being done by phone), as well as being mentally present. There are no interruptions or comments, not even non-verbal signals, instead they simply listen with empathy. The aim is to practise compassionate listening.
- After the five minutes, they switch roles and do the same exercise the other way around.
- The exercise is about listening with complete acceptance and no judgement, and not attempting to solve the other person's problem.
- To give someone our full attention without judging them is a powerful act of kindness.
- At the end of the exercise both people can discuss how they found this experience, if they wish.

Exercise 3: Swapping Places

- This exercise helps us to imagine being in another person's shoes.

- Imagine them sitting in front of you, facing you. Or you can practise this with another person, where you both do the exercise, physically seated opposite each other.
- Mentally swap places with the other person; imagine becoming them, and they become you.
- Now you're in their shoes, inhabiting their skin.
- Try to explore how it feels to be that person and what's really going on inside them. What kind of struggles are they going through?
- Next ask yourself how you might seem to them. You're looking at yourself through their eyes. Try to experience how you might seem from their perspective.
- At the end of the exercise swap back into being yourself, and spend a few moments looking at them (either literally or the imagined person), with a sense of new understanding and compassion.

RUBY'S EXERCISES

Mindfulness is being aware of your thoughts and feelings. When it comes to relationships, it's about noticing whether you're projecting these thoughts and feelings on to someone else. Be aware if you're throwing your luggage at them; they have enough of their own.

Relationships fall apart when there's no communication and I don't just mean discussing what the kids are doing, I mean when they stop asking what's going on in the other's life; then the connection is cut.

Always ask with curiosity, not interrogation, and don't just ask but listen closely when they answer. This is a skill that needs to be trained. We usually can't stay focused for more than a few seconds before distraction pulls us to the next thing.

If you listen to the other as if it's fresh and surprising news, that will spark your interest and interest creates rapport.

Exercise 1: Civilized Talk

- This exercise will teach you how to negotiate peacefully and productively.
- Only do this if both parties agree to do the exercise. This is a chance to state any needs you might have in the relationship. Usually, when we start expressing our needs, we get caught on a treadmill of accusations: 'It's your fault' followed by 'No, it's your fault.' (This can go on endlessly.)
- Before you start, sit for a few minutes and focus on your breathing. If you feel steady go for step one.
- In my opinion, unless you use mindfulness or some other training you won't be able to hear each other clearly when you get agitated because both your stress levels will hit the ceiling and then the 'blame game' takes hold.

Step 1. Be clear
One partner states what they feel they need from the relationship without interruption from the other. Take a few minutes to do this. (This is difficult because we're not taught or encouraged to state our needs clearly without feeling guilty.) They do this without blaming or accusing, they're simply saying how they feel. Be aware of tone; notice if a nag or a whine starts to creep in.

Step 2. Repeat back
For a few minutes, after the first person makes their statement, the other partner makes sure they've heard it correctly so as not to misinterpret what they said. They should always try to listen with new ears, not just here-it-comes-again ears.

For more clarity, they could ask, 'What I understood you said was . . . is that correct?'

This isn't a time to retaliate or slide into mudslinging. For example, 'You never listen to me.' 'I don't listen because you're always nagging.' 'I'm always nagging because you don't listen.' The idea is to just try and understand what the other person means. If you can't remember what they've said (it might be because we usually don't want to hear it) write down what they're saying.

Step 3. Using the 'I' word

Now, the listener responds with how they feel about what they've just heard. Don't use the accusatory 'you', as in 'You're always complaining about everything.' Stick with feelings, say, for example, 'When you say that, I feel helpless/guilty/angry/relieved . . .'

Step 4. Change places

Do the exercise again, changing roles: the listener becomes the speaker, the speaker the listener.

Step 5. Resolution

Discuss what you can each do to accommodate the other's needs, letting the ideas free fall. Ask each other what would it take emotionally, physically and financially to fulfil the needs of each other. If you both make a decision, discuss who's going to do what. If you both decide you'd like to go away, decide who's choosing the hotel. Who's making the reservations? Who's planning how to get there? Better to make the rules now before you fall into old habits.

Step 6. Reviewing

Tell each other how you felt about the exercise, trying to be positive. Say things like, 'I liked how you handled that when it started to get spiky,' rather than, 'For once, you held back from yapping.'

Note: If at any time during the exercise you find that things are falling into their usual rut or emotions are mounting, call time out. Agree on a time when you might want to try the exercise again, or seek a counsellor who knows how to hold the floor steady when partners have conflicts.

Exercise 2: Using Your Eyes as an Anchor

If you're with someone who's infuriating you, you can use your senses as an anchor to cool down your engines and stop you from retaliating or imploding. I use my sense of sight when I'm about to flip my lid. I'll focus my attention on a feature of the person's face: eyebrow, nostril (keep it near their eyes so they feel you're paying attention). If you focus on the shape, colour, texture, density of the eyebrow, you're in sensing mode and thoughts like, '*I hate this person . . . he's lying . . . she's cheating . . . I want a divorce* will fade. Whenever you feel you're about to blow your top, focus back to the eyebrow.

You could also do this by sending your focus to the sound of the other person's voice rather than the actual content of their words. Listen as you would to music or ambient sound; to the volume, tone, notes, quality and silences between notes. Stay with the noise, not the lyrics. The madder they get, the more it might sound like an opera or a construction site. If you can keep the atmosphere steady by focusing on sight or sound, in the end, they have nothing to bounce their anger off.

Exercise 3: Who am I?

Each of us is made up of many personas. My list is endless – what character I'm playing can change hundreds of times a day, depending on the situation. Here are a few labels I give some of my personalities:

The victim
The aggressor
The doer
The optimist
The pessimist
The exhausted one
The helpless one
The parent
The child

- Write down some of your personas. Don't analyse why you have them, just familiarize yourself with them. With the awareness that comes from mindfulness, you can tell which character you're playing at any particular moment. If you can observe your role, you won't be at its mercy, you'll be in charge. Once you know where your mind is, you can decide if you want to stay with that role or change it to someone more appropriate.
- Over time, it will become easier to recognize and accept your own and other people's negative and positive personas. The better you get to know yourself, the better you'll understand others. If, for example, they're in their devil mode, rather than switching on your devil mode, decide to switch to your adorable mode and then (hopefully) the other person will switch to theirs.
- I do this with Ed. If I'm in a bitchy state and I tell him, he knows to not tell me the boiler's broken and needs fixing. He knows I'll go into a rampage. If I tell him I'm in my nice persona, he knows it's an opportune time to give me bad news. I might even say, 'Boiler, schmoiler – who cares?'

THUBTEN'S EXERCISES

Exercise 1: Mindful Breathing

Breathing regulates stress and the state of the body and mind. It's difficult for kids to stay focused on their breathing, but here's an exercise they should find easy and, hopefully, fun. I would suggest avoiding breathing practices if the child has a health issue, unless under the right supervision.

Step 1. Ask the child to sit upright on their chair, like a king or queen sitting on their throne. Get them to feel where their body is in contact with the chair and where their feet are touching the floor.

Step 2. Ask them to take three deep breaths, breathing through their nose if they can (if it's uncomfortable, tell them to breathe through the mouth). Count out loud to guide them.

Suggest images to them for the out breath:

Picture the breath being like steam from a steam train
. . . or the breath of a dragon
. . . or strong wind blowing through the trees.

Also suggest that, when they're breathing out, they're releasing all their anxiety, worry and tension. (They don't need to use imagery for the in breaths.)

After the cycle of three deep breaths, they should return to normal breathing.

Step 3. The next step is for them to mentally count their breaths (in and out is one cycle) while breathing totally naturally and without effort. Ask them to breathe through their nose, but if it's uncomfortable then through the mouth. As they count each cycle, they can use their fingers – using the forefinger and thumb of the left hand to gently pinch the end of the thumb and then each of the finger tips, one by one, of the right hand.

Tell them to count, silently, for five or ten cycles. Whenever they get distracted, tell them not to get frustrated but simply to go back to number 1 and start again.

Repeat for a few minutes.

Step 4. Ask them to make a wish for happiness and peace for all. This is a little moment of compassion training. With all mindfulness practices, it's good to add a compassionate intention to the session. Kids can relate to this if it's very simple, like making a wish.

Exercise 2: Play Dough

This exercise helps young people to get in touch with mindfulness using physical sensations.

- Tell them to hold a lump of soft clay or putty (play dough is best, but anything that can be squeezed and moulded) in one or both hands. Tell them to squeeze it and to focus on all the sensations in their hands: the weight, texture and other qualities. When they become distracted, tell them to return their focus to the feeling of the clay or play dough in their hands.

Exercise 3: Grandmother's Footsteps

This game can be played with any number of participants. It's a well-known game, and it naturally encourages mindfulness, with a sense of fun.

- All the kids stand at one end of the garden except one kid who stands at the other end, facing away from the others. The kids walk very slowly and carefully towards the one with his/her back turned. That kid can turn around at any time, and then all the other kids have to freeze. If anyone is caught still moving, they have to go back to the start line. The winner is the one who manages to make it all the way up, and touch the back of the main kid. The winner then takes their place and the game starts again.
- This is mindfulness training through playing a game. As the kids are quietly creeping forward they have to be completely present and aware of their bodies, but also be totally focused on when the kid in front might turn around.

Exercise 4: What's the Time, Mr Wolf?

This one is a bit more intense, but kids love it!

- One kid is 'Mr Wolf'. They stand at one end of the garden with their back to the others, who stand at the other end in a straight line.
- As in the previous game, they creep forward, this time, chanting, in unison, 'What's the time, Mr Wolf?'
- Mr Wolf with their back turned says, for example, 'One o'clock' and the kids take one step forward; two steps for two o'clock, and so on. They're aiming to reach Mr Wolf, but at any time Mr Wolf can shout, 'Dinner time!' and chase the kids back to the start line. The first one to be caught becomes Mr Wolf and the game starts again.
- Some kids will take huge steps to try and reach Mr Wolf; others will take cautious little steps so they can run back to safety when Mr Wolf turns around. When the children notice this about themselves, it can help to nurture self-awareness and emotional intelligence.

- As with the previous exercise, begin the game by telling the kids how to be mindful while they walk.

Exercise 5: Exam Nerves

This exercise is good for exam nerves. It can be done any time before an exam or even for a few seconds during the exam.

Step 1. Sit in a chair and take three deep breaths, breathing in through the nose, out through the mouth. With each out breath, feel that all nerves and tension dissolve out of the body. Imagine the out breath as dark smoke, as if the body and mind are being cleansed of discomfort.

Step 2. Return to normal breathing and feel your shoulders drop and the stomach relax.

Step 3. Mentally count five or ten cycles of breathing. Breathe naturally, with no effort, and silently label each complete cycle; in and out breath is one, then in and out is two. If the mind becomes distracted, just go back to number 1. Do this a few times.

RUBY'S EXERCISES

If you're teaching a child mindfulness, pretend it's a game and, above all, NEVER BORE THEM!

From an early age you can start helping them sharpen up awareness of their senses. They don't need to know that awareness of senses lowers stress or that it creates neural rewiring for the better, just make it fun.

Exercise 1: Cloud-spotting

- Ask your child to think of their mind as a clear blue sky and their thoughts as clouds. The conditions of the clouds change, just as thoughts do; sometimes they're heavy, thunderous, dense, foggy, sometimes fabulously fluffy. Whatever the state of the clouds, the clear blue sky is always above them. If your kid doubts you, tell them to buy a plane ticket, go up there and check it out.

Exercise 2: Mouthing Off

- Another game that helps a kid to learn to focus on a sense is to ask them to describe what the food in their mouths tastes like right now. They'll love that you're breaking the rules, encouraging them to talk with their mouths full. They probably won't get many words out because they'll be laughing and, when you get someone to laugh, they're at their most open for learning. Encourage them to really get into the taste, the texture, the temperature, and so on.

Exercise 3: Breathing Buddy

- As adults, it's hard to sit and focus on your breath; for a child, it can be challenging and scary. On the other hand, the fastest way to learn to self-regulate emotions and thoughts is through steadying the breath. Breathing patterns reflect our internal states. With hard, fast breathing, we're usually in our fight-or-flight mode; with relaxed breathing, we're calm. Again, you don't need to mention anything about mindfulness, just make experimenting with different kinds of breathing into a game.
- Ask your child to lie on their back and put one of their favourite toys on their stomach (their 'breathing buddy'). Suggest they give their toy a ride by breathing in and out, and watch the toy move up and down. They may begin to notice that the toy becomes steadier as their breathing

gets calmer. The steadier the breath, the happier the toy. Not a bad image for them to carry around.

Exercise 4: Making Homework not Work

- I mentioned in the chapter on kids that not everyone learns best when sitting at a desk. Each of us is equipped differently, so we're all going to have individual learning techniques, and kids need to figure these out for themselves. (Schools don't usually help you find it.)
- If your child isn't learning anything during homework time, and this will be easy to identify, because they'll be slumped over with eyes shut, now is the time to experiment with what might ignite their curiosity. You'll be able to tell when it works, because their eyes will be open and lit up. When it comes to taking the test, if they're relaxed and happy when they studied, they'll be relaxed and happy when they recall the information. If, before an exam, you recall a happy memory, cortisol levels are reduced by about 15 per cent.

My Story

I remember my parents almost bolting me to the chair and forcing me to do homework. It was not a success. No matter how long they made me sit there, nothing would go into my head. I'm not going to lie to you, when I was around twelve, a few times I cheated in tests. We were being tested on the Declaration of Independence and nothing was sticking in my brain so I wrote it out on my upper thighs. This is probably why my body is so flexible to this day: from having to be able to do a backbend to read what's on the back of my

legs. I do not recommend these methods but, due to personal problems at home, I couldn't learn anything. Eventually, when I left home later in life, I no longer needed to resort to tattooing American history on my flesh.

- Have your kid figure out how and where they learn best. Make sure you tell them that everything is possible. If they need to study something, they need to do the reading and then go over and over it in their minds wherever they feel most relaxed: while swimming, skipping, dancing, singing it out loud, playing ping-pong, lying in bed, teaching their pet or teaching their toy.
- The most important thing to teach your child is emotional intelligence. They'll probably never have to repeat the Declaration of Independence ever again in their life, but they'll always need emotional intelligence. It ensures a longer, happier life, more friends, less freak-outs and helps create peace on earth. Is there a better Christmas present?

Exercise 5: Dress Up

- When your kid dresses up, which they all love to do, have them tell you what it's like to be in the other person's shoes. If they're trying on Daddy or Mommy's shoes, ask them to tell you what Daddy and Mommy are like inside, how they think and feel. This goes for dressing up as fairies, goblins, monsters, ghosts. This exercise teaches them empathy at an early age.

Exercise 6: Noticing

- Suggest to your child when they're walking down the street, watching YouTube, playing a video game or watching TV to imagine what the person or character on screen is like. You can turn the volume off, if it's on a screen, and have them imagine where the character lives, who their friends are, what makes them laugh, were they ever abducted into a UFO? This is early training for empathy, and that's the greatest training you could ever get or give anyone.

- If you get really good at this, you can even start imagining what people are thinking behind their tweets. Probably, *Help me, I'm lonely, please know I exist.*

THUBTEN'S EXERCISES

Through mindfulness we can gradually become less controlled by addictive impulses. They, too, are thoughts and, through training, we can learn to let go so that the addiction will have less of a hold over us. Mindfulness needs to be practised regularly, not just in emergencies. Just as regular exercise helps your body to retain less fat, regular mindfulness practice will get the mind to cling less to habitual patterns.

Working with the body scan, or the breath, as in the exercises for thoughts, are good ways of focusing the attention away from the craving, breaking the cycle of addiction.

Exercise 1: Investigate the Craving

- There are two ways to investigate: one is through directly focusing on sensation, the other is through constructive thinking.
- Sit upright in a chair in a quiet place, with good posture, or lie on the floor.
- Become aware of your body where it makes contact with the floor or the chair, just for a few moments.
- Method 1: Then focus directly on the feeling of addiction itself; maybe the craving feeling is already present as a sensation in the body. Don't chase the object of the craving, in fact drop all the stories and zoom in to the exact sensations. The feeling has now become your object of meditation, just as you would use body or breath in mindfulness practice.
- As you shine the light of awareness into the feeling, it may even start to dissipate and dissolve.

- Method 2: This is different from previous mindfulness exercises. It's called contemplative mindfulness, where you use your thoughts to gain insight or a different perspective.
- Gently ask yourself these questions without getting caught up in story lines.
    - 'If I feed this addiction will it just come back stronger?'
    - Or, 'Is there a loneliness or a sadness underneath this?'
    - Or maybe there are other questions you'd like to ask.
- If your thoughts start cycling too much, bring your focus back to the physical sensations in the body to ground yourself for a few moments before you go back to the investigation.
- When you ask the questions during mindfulness, you're coming from a deeper place; you've grounded yourself with body and breath, and you're now gently exploring, within a meditative state. Let insights arise as you compassionately investigate the craving.

Exercise 2: Your Higher Power

- Particularly in AA (or any other A's), people talk about their 'higher power'. For some people, this is something religious; for others, it's an energy; and some people don't know what it means but would like to feel it.
- Sit quietly, in a good posture, and focus on your body and breath for a few moments, to ground yourself in the present moment. Now visualize your higher power. Imagine this is above your head, or at a slight distance in the space in front of you. Bring up an image of whatever represents this source of help and support for you. It may be Christ, Buddha, or another religious figure, even the image of a ball of bright light filled with love and compassion.

- Mentally reach out to the higher power, let yourself open up, surrender and ask for help. In response, imagine that the higher power emanates rays of light or a healing nectar which fills you up completely, soothing the addiction and bringing relief and inner strength. At the end of the session, you and the higher power become inseparable; the higher power melts into light, which melts into you.

RUBY'S EXERCISES

Exercise 1: Noticing Novelty: Stop. Pause. Notice

- A large part of addiction is about seeking novelty. This is why the kick, or buzz, always has to be more intense than the last time.
- Pick up a drink, and try to notice five things around you you've never noticed before. It may be the colour of the liquid, the feeling of the seat you're sitting on, the type of glass you're holding, the music playing. This exercise will train you to see things as if for the first time, rekindling your sense of curiosity, and stop you falling into a state of agitation or boredom that ignites the cravings.
- If your drug is the internet, try to pause and notice the food stains on the screen (mine is covered in them) or the details of the screensaver to help you come off autopilot. This technique goes for any and all addictions: Stop. Pause. Notice.

Exercise 2: Dealing with Digital Addiction

- You can use technology mindfully, so that it works for you. Put your mobile in your hands or just in front of you. Notice when you want to reach out and have the urge to use the phone and try to sense where the feeling of the urge is in your body. Is it in your chest, your arms, your jaw? (That's where mine is.) Does the area of craving feel constricted, squeezed, burning, achy? (Mine is achy.) When you notice the yearning for your phone, don't beat yourself up; at least you noticed. The more you notice, the more successfully you'll be able to just make the calls you need and just send enough emails to keep on top of things without toppling into an overdose. The more you practise, the easier it will be to pull back from needing a phone fix or a hit of email.

My Story

Some of my hot spots when I get into my craving mode are my jaw, my shoulders and my chest. When I craved cigarettes, I'd always notice my jaw go rigid, like I was the Alien, about to tear someone's guts out. When I stopped I started chewing nicotine gum – for twelve years. I totted it up and realized I'd spent about £30,000 on chewing. That's what finally stopped me – the bank was empty because of chewing. I was so ashamed.

Exercise 3: Touch-typing

- This exercise trains you to notice when you're no longer producing anything useful or making sense – it signals it's time to take a break; and move away from your work for a limited time. Failing that, hire a great editor to fix your mistakes.
- While you're typing at your computer (one of my addictions, and which, coincidentally, I'm doing right now),

notice the urge in your body to push you to go faster or work longer when your brain is fried, as mine is now. Rest. When you feel the urge is no longer there, return to your work with a clearer, less driven mind.

My Story

So now as I type, I'm aware that I've lost the plot and my fingers are on autopilot. I can't stop because of the adrenaline and dopamine rush I'm getting. A minute ago I got a text from a friend asking if I want to take a walk. I wanted to but it's so hard to pull my fingers off the keys and my focus from the screen. I'm going to try to stop right now. 1-2-3 stop. I've stopped.

It's a few hours later . . .

Okay, I just came back from the walk. What's always amazing is that, as glued as I get to typing, as soon as I change the landscape (even getting up and facing another direction) the compulsion is completely gone until I sit down again, which I just did and now I'm typing again – but not so insanely.

Exercise 4: Shopping-mall Addiction

I made this next one up because of terrible experiences I've had in shopping malls. I may go in a mall for one small item but then get caught in an orgy of 'purchasitis'. I sometimes don't even like what I'm buying but the hunger won't stop until I have to be yanked out at closing time, dragging bags filled with things I will never wear. So for people like me here's my suggested exercise, it may work:

- Go to the mall, sit by one of the many burbling fountains and notice how many times you're overwhelmed by the urge to get up and shop. Notice how your head is pulled to various displays and the hunger to buy. See where that itch is in your body and locate the exact area that makes you feel you'll die if you don't buy.
- Now let that go and when the urge comes again, send your focus to the sound of the fountain. Do this over and over again until you can bear it no more and then catapult yourself into Zara.

Exercise 5: Breaking Food Addiction

- This exercise helps you to resist instant gratification and slows down your eating. Sit down with a large plate of food in front of you. Take a few breaths in through your nose, smelling the food. Now be aware of picking up the fork, feeling its weight and the sense of spearing the food. Feel the movement of lifting it to your mouth and your lips pulling the food off the fork. Now notice the details of what's in your mouth and the thoughts and feelings that come up with it. When your mouth is full, focus in on the texture, taste, chewability and maybe your urge for more will vanish.
- Notice if that urge increases as soon as you've swallowed to quickly cram in another mouthful even before the one in your mouth has gone down. At this point try to put down the fork and do some mindfulness for a few minutes, following your breath. Maybe the next time you eat, you can hold off ramming another forkful in or at least savour the taste of each mouthful, which makes you eat more slowly.

THUBTEN'S EXERCISES

The future is very much about technology and it, too, can be used for mindfulness training. People are often addicted to their phones, so why not get the phone to do something truly useful? It's all about using technology skilfully. There are mindfulness apps out there, such as the brilliant Headspace, and I'm building one called Samten. Also, Nick Begley, who was the former head of research at Headspace, runs 'Psychological Technologies'. He has created two apps: me@mybest and Rebalance with Mindfulness. We need a teacher to help us learn, so why not let technology be the teacher who starts you off? It's like using water wings to learn how to swim, or stabilizers on a bicycle: it trains us, but eventually we need to let go of these things and go solo.

Exercise 1: Not Getting Lost

- You can do this while watching TV or the computer screen. The exercise is to maintain a sense of the present moment, to feel that you're not getting too lost, or sucked into what you're watching. You're standing back a little, as an observer, but you're not getting distracted from what's happening on the screen. So, you're watching, but you're also aware it's an illusion.
- The easiest way to do this is to bring your attention to your body from time to time while you're watching TV or using the computer. You won't miss anything that's on the screen, it won't spoil your enjoyment; in fact, it will help you stay relaxed, focused and give you a sense of freedom while watching.

- It's brilliant training for learning to stand back and watch your thoughts and emotions without getting lost in them, because watching TV is like watching the mind.

Exercise 2: Social Media

- Try to use Facebook or any other form of social media as a practice of generosity rather than a search for validation. Often, we post our experiences on social media to see if other people like them before we even allow ourselves to like those things. We can end up becoming addicted to needing others to validate our experiences. We lose the ability to trust who we are.
- The exercise will help you to use social media consciously as a practice of giving – sharing rather than needing. Check your motivation before you post something online. Become aware: are you caught up in craving? Are you doing it to fuel a need for validation? Of course, it's okay to need something, but be aware of it and then the habit changes.
- This is all about using technology in a conscious way, so that you're running the technology, not the other way around.

Exercise 3: Future Thinking

- Often, we live in the future, constantly planning or worrying about 'the next thing'. But the next thing never truly arrives, because our mind has got into the habit of jumping ahead and so is already busy with the thing that comes after the next thing. We never get any rest. We never arrive.
- Begin by practising mindful breathing. Notice when your thoughts are racing into the future. Note this with a

mental label, saying to yourself, 'Future thinking'. Do the same with 'Past thinking'. Do you start to notice that pretty much all our thinking is about the past or the future? When are we ever truly in the present moment? Try not to berate yourself over this, just notice this pattern with curiosity and the habit can begin to change.

RUBY'S EXERCISES

Many of us already have technical add-ons and are still able to practise mindfulness. I have caps on some of my teeth and screws in my toes. This doesn't stop me from being able to send my attention to my foot, even though most of it is held together with screws and metal plates from my bunion operation. I might not feel the real toe made of my flesh but I still can send my awareness to the area, thereby buffing up my focus muscles in my brain, which has no screws and some loose screws, which is the point of the exercise. Let that be an inspiration: no matter what is artificial on or in your body, you can still do a body scan.

Exercise 1: Body Scan in the Future

- Even if, in the far future, you're just a brain floating in a jar, send your focus to where you imagine your toes were.
- No feeling?
- Then try your knees?
- Nothing? A nose?
- If you can't do that, see if you can focus on the jar that you're bobbing in.
- No? Okay, can you feel the formaldehyde surrounding you? Is it warm? Tingly? Wet? Anything?
- I'm not hearing anything. Okay, can you just sense if you're enjoying yourself or not in there?

- Well, don't get depressed, even if you can't hear me, you're still you and that's all that counts.

Exercise 2: Teaching Your Child Compassion in the Future

- When your kid is wearing his Oculus Rift glasses, and he's interacting with some virtual-reality creature, ask him to try to feel what it would be like to be in their shoes, even if they're shaped like a lobster. (Virtual-reality characters tend not to resemble people.) If your kid senses the lobster is suffering and feels they'd like to do something to make the lobster feel less alone, maybe sit with him like a close friend, then he is learning the golden rules of compassion. Even a lobster needs love and you should be proud.
- I can't think of any other mindfulness exercises for the future because it's not here and I'm too present so forgive me.

12

Forgiveness

I planned to call the last chapter in this book 'Forgiveness' because to forgive is the most difficult of all human endeavours. One expert on compassion says, 'Forgiveness is the solvent that dissolves the glue that holds our self-righteousness tightly. It softens the feeling of "I am right and you are wrong." When I am able to forgive, I recognize the humanity in others.'

This forgiveness thing is a tough one because it goes against our primitive nature. If we feel that we've been wronged or something is unfair, it's embedded in every cell of our body to seek revenge. In every film, book or play you'll ever see with a bad guy in it, people will want him hurt, and hurt bad. Phrases like, 'Hang 'em high' and 'I'll be back' ring in our ears far longer than 'Have a nice day.' I was getting nowhere trying to write this chapter because 'forgiveness' is not really in my CV. Coincidentally (which is how my life runs), at around the same time I got a phone call from the makers of *Who Do You Think You Are?* During the making of the show, I learned what forgiveness was.

Who Do You Think You Are?

After my mother died, I found an old leather suitcase in the attic that she must have brought to America with her when she escaped from Vienna. It was full of hundreds of letters and photos; I had no idea who anyone in them was. My mother never mentioned any relatives or the past, so I assumed I'd just dropped from another planet. I knew we had distant relatives somewhere, because they were the ones who got my mother out of Austria, but there was never any mention of direct ones. I gave the TV researchers the suitcase and, eight months later, they said they had traced my past and I was booked.

At the initial meeting with the team of *Who Do You Think You Are?* I was asked what I wanted to find out? Well, for a start, I wanted to know why my parents were the way they were. Was it because of the war, or would they both have had mental problems anyway? I also wanted to know why I've always felt so haunted; even though my parents never said a word about what happened, I used to wake up most mornings with a sense that World War Three was about to break out. Our doorbell sounded like an air-raid siren and I'd come down with my hands up in surrender mode every time it rang. As a kid, I built up rations to hide in my homemade air-raid shelter in our basement. I collected cherries from trees, and canned food, to fill my bunker. I thought the Russians were coming over the horizon to kill us all. In reality, we lived on Lake Michigan and Wisconsin was on the other side. I used binoculars every day to spot battleships.

With this kind of inbuilt paranoia, I'd want to know if it was my imagination or were my parents as weird as I thought. They told me I was a fantasist and I was the one that was mad; their job was to straighten me out. I wrote about them in my first book *How Do You Want Me*, which Carrie Fisher helped edit. She said it was almost as dark as her background and, in the darkness stakes, that's as complimentary as you can get. On the other hand, they did boost my career with the great material I got from their fantastically surreal lines, which I didn't even have to edit, I wrote down exactly what came out of their mouths and used it for many of my shows. Our dog once ate a small sock of mine; my mother waited until it came out his other end, washed it and put it back in my drawer. When I asked why, she belted, 'People are starving in Bavaria.' My father, equally strange, wrote his will which I found when he was very old. It said I'd only get 35 per cent of what he was leaving when I turned forty-five because he thought I'd be insane by fifty or a heroin addict. By the time I was seventy-five I'd finally get the rest of the cash. He used to say, 'Who's going to marry you? Your behind is as big as a house?' I think he was subtly informing me he didn't hold out much hope for me. His pet name for me was 'sad sack'. He once said, 'Of all the millions of people, why would they choose Ruby Wax?' This was his response when I told him I had got a job in a mini-series for NBC in America. He later told me, 'I called them and they said they never heard of you.' He had called the reception desk at NBC.

It felt like they were bringing the Second World War into our kitchen. Every day the battle lines were drawn in the breakfast nook and we'd lob lethal verbal grenades at each other. With my background, I could only have ended up as a criminal or a comedian. But no one has cracked the whip to succeed as hard as me on me; I probably did it to prove to my father, even after his death, that I wasn't the failure he had predicted I was going to be.

In a sense, my parents were my inspiration and my hyperdrive to make good. So I think you can see why I was curious to find out how they ended up the way they did and who they really were.

Day 1: 28 June

I flew with the crew to Vienna; both my parents had lived there. We got to the hotel, which had a gulag decor and a prison warden at the desk who hated me because I asked for a room that didn't have the sound of a motorway running through it. The bed was a hard slab of wood with a thin-tissue duvet. I spent my first night googling other hotels and moved out in the morning. I told the crew that, because this journey might be harrowing, I needed room service and a pillow.

Day 2: 29 June

Today, the director handed me photos of my mother, who looked like a movie star, doing movie-star poses. Next to her Greta Garbo looked like an old Chihuahua. In many photos, she was with different men, who I assumed were boyfriends: skiing with them in the Alps, lying on a beach with them in what looked like the Riviera, always looking chic and always unbelievably beautiful.

For lunch, we went to one of those dark-wood-walled Alpine Hofmeister beerkellers; decor – very early Nazi. There were paintings on the walls of people (I kid you not) having oral sex with smiling angels looking down from above. You look at these while you're eating schnitzel.

You know how, when you get to certain places, nothing seems to make sense? This city is like that. Later, we walked by a toy shop and there was a window full of children's teddy bears, but arranged in orgy positions.

At the Hofmeister Keller I was introduced to Eleanor, a historian, who walked me around the corner and told me that this was where my father had been in prison. He had told me that, as a young man, he'd been in jail but led everyone in aerobics to keep them fit. Eleanor informed me that he was put in jail for being a Jew and that he hadn't taught aerobics; the inmates were being tortured. The Nazis were making them all jump like rabbits; the older ones would fall over and the guards would beat and humiliate them. She said that, maybe because my father was young, he'd been able to keep up with the jumping. Then she handed me a postcard my mother had written to my father while he was in prison. My mother told him to be careful and brave because he hadn't done anything wrong and that she cried each day, thinking about where he was. She wrote that as soon as he came out she would have 'nice food' for him. That almost killed me. I didn't know she ever loved him, and the expression 'nice food' was so pure and innocent. There were other postcards from her that seemed to be about trivia – but they were in fact plans for him to escape, written in code.

Eleanor showed me the original letter from the Gestapo informing my father that he would now be under their 'protection' – by 'protection', they meant they were going to send him to Dachau unless he left Vienna after prison. At this point, they just wanted the Jews out. The exterminations happened later. But, almost overnight, the Austrians turned into savages; people who were friends and neighbours started to beat up Jews in the streets and rob their homes. The police turned a blind eye, wearing their swastika armbands and allowing the massacre to rip. The German SS wrote to the Austrian citizens saying that they shouldn't be so hard on the Jews, to hold back on the beatings, telling them that, if the Jews were going to be punished, it would be done according to German law. The world would soon find out what German law meant.

Day 3: 30 June

I was taken to the flat where my father and his mother lived in the thirties. It had impressively large rooms, high ceilings and overlooked the canal. Clearly, the job of 'intestine merchant' was a money-spinner. Maybe I forgot to mention it, but that was once my father's job description and he took that talent with him to the US. (I always tried to glam it up, saying he was a fashion designer for hot dogs.) Anyway, there was definitely money in guts. In 1938, the Nazis knocked on the door and took him off to jail. When I was in the house, I looked out of the back window to see if he could have made an escape, but it was too high up. The historian I met there told me how my father had escaped to America.

When he got out of jail, after the Gestapo kindly told him, for his 'protection', to leave Vienna immediately or go directly to Dachau, my father bought a plane ticket to Belgium. In 1938, the ticket cost the equivalent of £2,000 today. So where did he get all that money at a time when it was rare for even the rich to fly? No one knows. When he arrived in Belgium, they wouldn't let him in until he could prove he had money in a Belgian bank to show that he could support himself. He did. How he managed to get money into a foreign bank is also a mystery. And the biggest jaw-dropper of all was that, with only two weeks to go before the Nazis took over Belgium and started rounding up the Jews, my father stowed away on a ship going to New York. This, the historian said, was a one-in-a-million shot. How did he get on the ship, and how did he get off, when it was impossible to get into America without proper papers and at a time when America was no longer allowing entry to Jews?

It seems my father sold guts and had guts. I'm very proud, because it seems the rest of the family had the same chutzpah. His brother Martin and their mother got to Cuba and then America. His other brother, Karl, went to Martinique and then to the US. His sister went to London and then to the US. His father deserted the family when he was young, so no one knows what happened to him.

Before my father left for Belgium, he quickly married my mother, who was twenty-four at the time. I have the photo. They looked like a couple straight out of a film. Another specialist took me to the synagogue where they married. The synagogue was no longer there, it was just an empty space between buildings, like a missing tooth. I learned something else I wasn't aware of; my mother was in Austria for Kristallnacht. It's usually translated as 'The Night of Breaking Glass', a night when the Nazis went wild, burning down thousands of Jewish homes and most of the synagogues. Luckily for her, shortly afterwards she received a letter from her distant relatives the Hambourgers and, a month after Kristallnacht, she set sail to Chicago. I'm pretty sure now what would have helped throw her over the edge was that, at such a young age, she was prematurely evacuated from this city of wall-to-wall splendour: museums, opera houses, theatres, cafés so ornately decorated it seems every building was sculpted out of whipped cream. She was blonde and blue-eyed, a serious babe with a highly educated brain; most probably an 'it' girl in Vienna.

If my mother had just mentioned what had happened to her and how she felt, I would have understood and forgiven her for her hysterical fits of rage. Maybe she was trying to protect me. Maybe she was too traumatized. Or maybe she simply didn't think I had the capacity for compassion.

Day 4: 1 July

In the morning, we stood outside the door where my mother's aunt, Gabriele (her nickname was Ella), and her husband, Salomon, had lived. (I had no idea I had a great-aunt.) I was met by a historian called Doron, an expert on the Jewish community in Vienna in the thirties and forties. A sign on the door read 'Dentist'. Doron told me that Ella and her husband had also been dentists and showed me letters addressing my mother as 'Golden Bertal' and sending a thousand kisses, thanking her for her affidavit from America (my mother had already escaped). Doron also told me that when a friend or a neighbour would come to a house and say they were going on a 'long trip', everyone knew what it meant. They meant they were going to commit suicide. No one tried to stop them. If they did commit suicide, the family who were left were punished.

Ella's last letter is dated 8 October and said that the situation is now life or death. Each day, a thousand people were rounded up from the neighbourhood and taken off in trucks and trains. It seems that, although my mother sent an affidavit sponsoring my great-aunt and -uncle, the laws were constantly changing and, by 23 October, Austria had been annexed and no one could leave. I asked Doron what he thought had happened.

My great-aunt Ella and great-uncle Salomon were in their sixties when they were deported. Doron wouldn't tell me what happened to Salomon and Ella but said I'd find out the next day; it was like a macabre cliff-hanger. After we filmed the scene, I wanted to switch my mind to something else, so I went online shopping and didn't get offline until 4 a.m. I didn't stop until I had purchased almost every shoe in a size thirty-eight from around the globe, even ones I didn't want.

Day 5: 2 July

I was put on a train to somewhere; as usual, they wouldn't tell me where. I kept losing my temper because the Wi-Fi didn't work. In the back of my mind, I was disgusted at myself, imagining people on a similar train years ago, travelling to unimaginable horror, beating on the doors to get off. We got off in Prague. I was delirious with joy, as I had always wanted to go there. The streets were lined with colossal homes; three hundred years ago, each neighbour trying to out-turret, out-mosaic, out-Greek-statue the next. The multitude of churches also competed for how much bling could be crammed into a square inch. I couldn't stop thinking that, if Jesus died for our sins, he shouldn't have bothered. We were and still are sinners. And one of the proofs of how sinful we are is our habit of overdosing churches with more wealth than could feed the world thrice over. I couldn't stop thinking about that speech Hamlet does in the play about him. (I love this speech).

> What a piece of work is a man! How noble in reason, how infinite in faculty . . . ! In form and moving how express and admirable! . . . And yet, to me, what is this quintessence of dust? Man delights not me.

Day 6: 3 July

The following day, the smile I had from seeing Prague was wiped right off my face when I arrived in the town of Theresienstadt, an hour away. I had told the director I wasn't going to any camps, if camps turned out to be involved in my story. I was told this wasn't a camp, it was a ghetto. When we got off at the train station a chill went down my spine when I realized that my great-aunt and great-uncle would also have disembarked here; these were the tracks that had brought them here. I was told they were marched two kilometres through the snow to the ghetto, with hardly any clothes on, and beaten as they walked. The ghetto now looks like a quaint village: multi-coloured houses and tree-lined, cobblestoned streets; a soundtrack of chirping birds.

At the start, the Nazis allowed the Jews to put on their own entertainments in the ghetto – operas, plays, even a band that played 'Ghetto swing'. The elderly were treated the worst because they were of no use so they were literally sardined into the attics with no windows or ventilation. And here's what will give me nightmares for ever – I was told there were no toilets and diarrhoea was rampant. My relatives died early, which was lucky; Salomon after a week, Ella probably earlier.

Day 7: 4 July

A day travelling from Prague to Vienna and, again on the train, I couldn't stop thinking about the tracks that had carried the Jewish cattle cars.

Day 8: 5 July

A day off filming. I visited my mother's house, now a small food shop run by a Muslim. Oh, the irony. The new scapegoats.

Day 9: 6 July

Breakfast consisted of something fishy (I think it was fish, could have been beef cheeks) floating in kerosene, with an accompanying bowl of icing sugar. After this scrumptious meal, I was taken to a graveyard and given a map to find the graves of my grandfather and great-grandfather, Richard and Salomon. I don't know why, but when I found where they were buried, some basic instinct made me start stroking the grave. It was a pitiful gesture to somehow get close to these people I hadn't known existed. When the cemetery first opened, it was too far out of town and no one wanted their loved ones buried there. What did they do? They dug up the graves of Beethoven, Strauss and other celebrities of the day from their hallowed ground in the centre of Vienna and reburied them in this out-of-town graveyard. It then became 'the' place to be buried – the hottest ticket in town.

I met Chaim, who leads tours through the graveyard. He walked me to an unmarked plot and told me it was where Ella's sister, my other great-aunt, Olga, was buried. There was no tombstone to identify her. He explained that this was how they buried the penniless in those days. The social services brought them here, but the family had to buy the stone. It broke my heart to stand beside this unmarked rectangle of dirt. A question in the back of my mind was why didn't her family get her a stone?

In the afternoon, I met a historian, Sabine, who asked me to guess what had happened to Olga. I asked in a desperate voice if she had been an actress, which is what I was hoping. For some reason, I was convinced that someone in my family had been and I had high hopes that it was Olga (she looked a little like me in the photo). Sabine handed me a photocopy of a clipping from a Vienna newspaper and pointed to an article that said Olga was suffering from either 'idiocy', or madness, and had to be taken to an insane asylum. That was not the news I expected. I asked, in a tiny voice, 'How long for?' 'Thirty years.'

She brought out the original leather registration book, about three inches thick, which contained fastidious handwritten lists of inmates, the number of visitors and dates of death. One thing you can say about the Germans and Austrians, they know how to keep meticulous records. Olga died in 1938, of tuberculosis. My mother was nineteen when Olga died, so she must have been a witness to Olga's 'idiocy', or madness. Back then, they didn't have specific names for mental illnesses so they labelled chronic cases as 'agitated'. She must have been highly 'agitated', if her stay lasted thirty years. I asked if she had been an actress before the 'agitation', thinking maybe the two didn't cancel each other out. 'No,' I was told, 'she was a seamstress.' I still hoped she could have part-timed as an actress when she wasn't seaming.

From there, they took me to the 'insane asylum', Steinhof, where Olga had resided. We drove down a long drive surrounded by endless bucolic lawns, dotted with fountains and enormous leafy trees. There were about sixty palatial buildings, 'agitated' patients shuffling between them. Clearly, it was still open for business. The director of the show thought the asylum would upset me, but I felt like I was home. I've always loved institutions because I always feel safe with my people, and now I was surprised to discover they weren't just my people, they were my family.

The buildings were lined up in rows. The bottom row was for the 'quiet' inmates, the next for the 'semi-agitated' and the last, at the top of the hill, was for the 'agitated'. This was where Olga must have lived. During her stay, there were up to five thousand patients living there.

Sabine explained that the therapies offered in the days of not knowing anything and being totally ignorant were things like the lukewarm-water therapy. This is where the patient is left in a bath for days on end. (Obviously, it never worked, but everyone was very clean.) Another was sleep therapy. Simple and yet not safe. The patient was tranquilized, put to sleep, probably to 'sleep off' the mental illness.

When the Nazis arrived in Austria with their Final Solution, they came to this asylum to experiment on the inmates. They considered the mad 'lower than dogs', so they didn't hold back. They performed experimental surgery on the kids and tried new and effective ways of using gas on the adults. By this point, Sabine didn't want to continue. Luckily, Olga died before all that happened.

Day 10: 7 July

We went to the Czech Republic, to a city called Brno. I'm going to give it nil points for charm. I went to a building that held the archives for the area. (I'm now so used to archives, I can start doing reviews, comparing the decor of the filing cabinets and their dust.) The archivist slid a very fat, weather-beaten journal under my nose, opened the book and pointed to the name Berta Goldmann, my great-great-aunt. My mother's name is Berta so she must have been named after her. I asked, again in my desperate, sad voice, 'Was she an actress?' I was told to read out the name on the book cover (she gave me a translation). It read, 'Brno Insane Asylum 1883–1902'. I thought, *Is this kismet that I end up working in mental health, fighting the stigma of insanity, and now it appears my family tree is rife with it?*

It's strange that I was so fascinated by mental illness, starting around age thirteen, when I had no idea that either my mother or I were ill. I still have a library book from Evanston High School on a shelf in London that should have been returned in the sixties called *This is Mental Illness*. I must owe over a million dollars in late fees. How did I know that someday I'd have to face not only my demons but most of my family's?

The archivist then produced another Viennese newspaper clipping stating that Berta Goldmann 'sold all the furniture in her flat and left a note saying she was going to commit suicide'. They must have found her and taken her to the nearest asylum. It appears she was in for less time than her daughter, only seven months, and then she too died of tuberculosis.

So, off we went to visit the Brno asylum, down a long drive leading to a magnificent neoclassical building, painted a soft yellow, surrounded by undulating emerald lawns and burbling fountains. It's also, like Olga's residence, a working mental institution, now holding about one thousand patients, but I was told by my director that we couldn't film any of them so I took off to the cafeteria to mingle.

Language is not a barrier when bonding with my fellow mentally ill people. A girl, Eva, with bipolar, and I became inseparable. She spoke enough English for me to understand that she was on a manic high; that's why she was walking with ski poles: she was moving so fast she couldn't keep up with herself. Everything is free here and the nursing staff seemed kind and attentive. How bizarre that, compared to the UK, where there are hardly any beds for mentally ill people who need help, here, they have thousands. The buildings are impressive. In the early 1900s, the Austrian government wanted to show the world how well they could care for their 'insane'. That was before most of the population of Austria and Germany became more insane than anyone who ever walked these corridors.

I actually felt relieved to know that Berta had ended up here after her suicide note was found. I don't know why, but

I felt proud of Berta and Olga. Olga had no tombstone, and no one knew where Berta was buried. I lay on the lawn and decided I would buy a tombstone for Olga and Berta and engrave the words, 'They were great women. I'm very proud of them.' I thought they were probably wonderful women and that, in the right circumstances, they could have been actresses. I just lay face down on the grass. I didn't want to leave.

Day 11: 8 July

I woke feeling calmer than ever. My body and mind felt easy; not the usual tearing pressure to get up and accomplish something – anything – urgently. I could have saved myself a lot of agony by skipping trips to the therapist and going straight to genealogy.

The Monk, the Neuroscientist and Me

Ruby: We've talked about compassion, which I can understand, because I can potentially do something about the suffering of others. But what about when someone screws you over or hurts you? My instinct is always to kill them.

Neuroscientist: We're hard-wired to want fairness. If someone cuts you off in traffic and a minute later you see the cops pull them over for speeding, your nucleus accumbens, the brain's reward centre, gives you a buzz of pleasure. But if they get away with it, your anterior insula will become more active, making you feel physically uncomfortable. That biological drive for fairness helps people get along in society, but it can become toxic when it becomes a desire for vengeance.

Ruby: I know that feeling of revenge; it's reptilian, when I feel it, I turn into the Alien, I want to tear them apart. And this is even when I get a traffic ticket. I don't know where forgiveness comes into that.

Monk: I think we worry that forgiveness means we're letting someone get away with something, but it's more to do with releasing ourselves from the burden of resentment. Ruby, how does it feel when you're holding on to that anger, when you keep chewing over how unfair it is?

Ruby: Sometimes the chewing over is the best part.

Monk: But it's like holding on to a hot coal. It's you that suffers, not the other person. Forgiveness would help you drop the sense of burning.

Neuroscientist: Vengeance may be sweet, but anger and fear are potent activators of the limbic system. That's the more primitive part of the brain which, as we've said before, activates the fight-or-flight response. And holding on to anger or fear is a chronic stress. It promotes the production of poisons like cortisol; it's terrible for the brain and body over the long term.

Ruby: I can't believe you can train forgiveness. It seems so against most people's nature. What happened to 'An eye for an eye . . . '?

Monk: The problem with an eye for an eye is that everyone ends up blind; the cycle of revenge never ends. In mindfulness training, there's a step-by-step approach. The first step is to recognize that anger is a toxin and that forgiveness reduces it. The second step is realizing that your enemy is giving you an opportunity to develop a skill. It's like this person is putting weights on your barbell, helping you get bigger muscles. They're an ally in your mindfulness development. The third step is understanding the pain that's driving the other person's action. This involves developing compassion and wisdom.

Ruby: So, what happens in your brain when you decide to forgive someone?

Neuroscientist: The brain has to do two things in forgiveness. First, it has to temporarily suppress the feeling of anger that's happening mainly in the right dorsolateral prefrontal cortex, the brain region that puts the brakes on cognition. Then there's a network between the frontal cortex and the parietal cortex that updates your perceptions. That allows you to let go of your rigid view of blame, and the anger and stress can dissipate.

Ruby: I always have a motto – I should wear the T-shirt – 'Who can I blame?'

Monk: There's a parable about that. Someone throws a stone at you. Who do you blame? You blame the person, but it was actually the stone that hit you. So why not blame the stone? Because the stone had no intention to hurt you. By that logic, you should also not blame the person but blame the anger and suffering that made them throw the stone. When someone hurts us, we always think they're out to get us and that it was deliberate. But if you understand the mechanisms of the human mind, you know that people lose control and do and say things they don't intend when they're in pain.

Ruby: What happens if the person really is an asshole and deserves to be punished?

Neuroscientist: Maybe the guy does deserve to be punished, but your desire to punish him will make you miserable. As long as your limbic cortex is hyperactive, your stress levels will be high.

Monk: Three years ago, my teacher and closest friend was brutally murdered while on a trip to China. The killer had been a monk in our monastery. We all knew him. I guess it was a moment of insane rage; none of us can understand why he did it. Before I became a monk, twenty-five years ago, I would probably have been filled with hatred and vengeance, wanting to hunt the killer down. But when it happened, I felt shock and grief but no trace of anger or revenge. It's clear to me that the guy needs to be locked up because he's dangerous, but instead of hatred I just felt sad and concerned. I found myself worrying about how he was doing in prison and what would happen to him. I think this attitude was a natural result of all the practice I've done.

Ruby: That's really heavy. I don't know if I could ever do that. When I was in Austria doing *Who Do You Think You Are?*, I knew there was anti-Semitism before the war, but that people basically lived in peace with each other. If you were Jewish, you could still go to university and become a doctor or a lawyer. Then, almost overnight, your neighbour, your friend, your co-worker, turned on you. Not just stopped speaking to you but went feral; raped and beat you and didn't stop the trains that took you to your death. The Jews hadn't killed anyone, so it wasn't revenge. Like Rwanda or Bosnia, the ethnic minority hadn't murdered anyone, so why did they get annihilated? Ash, what happens in the brain that turns us into savages?

Neuroscientist: That's the question that everyone was asking after the Second World War. A psychologist at Yale, Dr Stanley Milgram, did pioneering work on this in the early sixties. Milgram wanted to understand the psychological defences that the Nazi leadership used during the Nuremberg trials, mainly that they were just obeying orders. Milgram placed an ad in the local paper recruiting volunteers to participate in an experiment that he told them was about learning. The volunteers were given a list of maths questions and answers and each one was told to play the part of a teacher, quizzing another volunteer in a closed booth. When the person in the booth got a question wrong, the volunteer/teacher was supposed to press a button administering an electric shock. The shocks increased in voltage from slight to severe to life-threatening. It was all fake: the guy in the booth was an actor and there were actually no electric shocks. But Milgram found that every single one of his volunteers continued delivering what they thought were high-voltage electric shocks, even though they heard the other guy screaming in pain and begging to be let out. About a third of the volunteers even delivered what they believed would be life-threatening shocks. They obeyed because each time they hesitated in upping the voltage, Milgram would say, 'Please continue.'

Ruby: And with 'Please continue,' that did it? I'm not buying this.

Neuroscientist: The experiment has been replicated many times, in many countries, over many years, and there have been similar results. What seemed to influence the original volunteers to continue was that Milgram told them that he would take full responsibility. So even though the volunteers were pressing the button, they didn't feel it was their fault. At Nuremberg, Nazi soldiers said the same thing: they were just following orders, it wasn't their fault.

Ruby: The Austrians turned on the Jews without any instructions. The Germans even told the Austrians not to be so brutal when attacking the Jews. They wanted to take charge.

Neuroscientist: So that's something on top of obedience to authority. We're tribal animals, so we define people as either in group or out group. We don't see out-group people as fully human, and that enables us to treat them in horrible ways. The important thing is that this can happen to any of us, at any time, if the circumstances are right. When an authority figure whips up enough hatred for the out group, it pulls together the in group and incites them to violence. And it's not all at once; usually, it's one little step at a time. It's like when people start a sentence with, 'I'm not a racist, but . . .', you know what they're about to say will be racist. It's the start of losing your moral compass.

Monk: When we can face the fact that we all have this inner savagery, we can stop demonizing the 'other', acknowledging that we all carry the seeds of violence and hatred within us. If we can learn to look inside and forgive ourselves for this very dark but human feature, we can begin to forgive others. That's where mindfulness helps: we notice our animalistic side but still manage to find forgiveness, when we realize that we're bigger than our thoughts.

Mindfulness Exercises for Forgiveness

If you can let go of endlessly thinking about why, what and how things have gone wrong for you and whose fault it is, then you've forgiven. And if you've forgiven, you're free.

THUBTEN'S EXERCISES

Exercise 1: Three Steps

- Forgiveness can be trained by using a three-step process. These steps involve thinking differently about a situation so that the anger and hurt can begin to shift and forgiveness start to emerge.
- You'll notice that these steps involve thinking, rather than the usual style of mindfulness, where you try not to get involved in thoughts; however, they're powerful tools for transforming our deep-seated habits.
- Begin by calling to mind the situation or person you are finding challenging. Your mind might start to swirl with resentment, but try to step back a little and ask yourself some questions about the situation:

1. **Who's Really Hurting?**
   The first step is to recognize that holding on to anger just makes us burn. This person who hurt us did so in the past, but what is hurting us now in this moment? Our resentment and anger. Holding on to this anger is like holding a hot coal in our hand, it just burns us. If we could drop it, we would find relief. In the same way, if we work on our minds, we can free ourselves.

2. **Gratitude**

Is our enemy really an enemy? They're actually giving us an opportunity to develop the skill of forgiveness. They're an ally in our compassion training. Without this person, how else would we learn forgiveness?

When you can think in this way, the idea of 'enemy' starts to change, as gratitude arises. Maybe your enemy is a friend in disguise.

3. **Understanding**

Who else is hurting? Think about the other person's deep pain and suffering. Maybe it's hidden, but we can be sure it's there under the surface. Try to resonate with the struggles which might lie beneath their aggression. Even if they're cold-hearted or they seem to enjoy causing harm, we could understand them as being unwell, completely out of balance, and that's what makes them behave like this.

This is a challenging step, but the more we practise mindfulness, the more we'll see how the human mind can so easily be controlled by thoughts and emotions; there's very little freedom until somebody trains their mind. With mindfulness, we start to understand the human condition. Nobody is 'out to get us', they're simply consumed by their own pain and ignorance and very often can't control their words or actions. This doesn't mean we're condoning what they do, or allowing it to continue, it's about shifting our attitude of resentment and dropping the burden of hurt and rage.

Exercise 2: Mindfulness and Forgiveness

The second method, to be practised alongside this, is simply to do regular mindfulness practice, using whichever technique you like, for example the breath. Through this ongoing training, our patterns of resentment, as with all negative emotions, will start to have less hold over us. We can learn not to latch on to the feelings so much. That persistent hurt is yet another habitual pattern in the mind, and mindfulness practice helps to loosen up those habits so they can start to dissolve.

It's important to be very patient with yourself and not to berate yourself for not being able to let go. Just do the practice, and things will slowly start to shift.

Exercise 3: Training throughout the Day

In the exercises for body, we looked at practising micro-moments of mindfulness throughout the day, particularly when we're waiting for something. The ability to be mindful and stay relaxed in queues and traffic jams helps us to train ourselves in forgiveness, as it teaches us to welcome difficult situations and not push them away. It trains us to stay mindful in situations where we would normally feel resentful, irritated or upset. This training will help us to see our difficult relationships, or the things we find hard to forgive, in a new light.

Exercise 4: Other Methods

Please also see the exercises on compassion and relationships, as they also work for forgiveness.

Exercise 5: Forgiving Ourselves

- Sometimes it's much harder to forgive ourselves than it is to forgive others. We blame ourselves, mentally beating ourselves up.

- If we can have some compassion for ourselves, this will really help. In the compassion exercises, there are methods for training in self-forgiveness.
- Also, one thing I find useful is to look at a photo of yourself. Sit quietly and closely examine the photo, especially the eyes and around the mouth. There you might detect some vulnerability, or some child-like, innocent qualities, and you can start to feel compassion for this person who is just doing their best, and is lovable and worthy of forgiveness.
- Self-forgiveness becomes easier the more we practise mindfulness, as the training helps us to see that our mind is bigger than our faults; those things come and go and are not our true nature.

RUBY'S EXERCISES

I can't think of a better exercise for forgiveness than getting booked on *Who Do You Think You Are?*

My Final Thoughts

I'm glad I did *Who Do You Think You Are?* It explained why, as far back as I could remember, I'd wake up hearing screaming in my head, and not just one person but a large chorus of maniacs. I'd feel that heart-stopping panic and fear even when nothing on the outside had instigated it. I could be on holiday, sunbathing on a lounger, and still hear it. The word 'insane' was bandied around the house when I was growing up. I remember my father telling me he thought I'd end up in an institution by the time I was fifty. I also heard it when my mother, during our fights in public, would sometimes grab a passer-by and belt out, 'Do you think I'm insane? Because my daughter thinks I am.' I'm starting to think that this 'agitation' my great-aunt and great-great-aunt had is alive and living in every cell in my body. It's woven into my DNA. I'm not in an insane asylum, I'm carrying one around in my mind.

The fact that my father had a drive to survive at all costs explains why I always feel this drive to take on any challenge presented to me. It can be something as trivial as jumping a queue (at which I'm a master) to pulling off getting into the Royal Shakespeare Company, writing television shows for the BBC, writing four books, or studying at Oxford.

I know that if I try to bury or ignore the battle in my head, it will just create more war. I feel like I'm on a railway line that splits; one track goes towards the insane asylum, the other to freedom. (Railway lines have haunted me since doing the programme.) I'm sure what propelled me to study mindfulness was to try to resolve my dysfunctional links, to be able to find some peace in my internal bedlam. Finding out about neuroplasticity, discovering that I can remap my inherited neural wiring and rewire it for a better life, was the greatest news I've ever received.

I came back to the UK on a Sunday night and that Monday morning went to collect an honorary doctorate for my work in mental health. I stood there in my red Harry Potter gown and square hat with tassel, giving a speech to about five hundred mental-health nurses. The Friday of the same week, I went as a visiting professor to the University of Surrey to give out diplomas to the graduating class. They 'robed' me up in my old Oxford graduation gown and, as I wedding-marched down the aisle and on to the stage, I thought about my parents, about Olga (my great-aunt), Ella (the other great-aunt), Berta (my great great-aunt), Richard (my grandfather), Salomon (my great-uncle), Karl (my uncle), Martin (my uncle) and all the others I'll never know, and wondered whether, if they saw this, they'd be thinking, *How the hell did that happen?* I'm sure they'd be confused, but maybe they'd be proud. That was a moment of self-forgiveness when I realized what I had done was against all odds, a million-to-one chance, just like my father escaping from Austria.

I also forgive my parents. Who knows who they would have been if they hadn't had to run for their lives? The past was not their fault.

There's no question that we humans are flawed, but the power of our minds can change everything.

Acknowledgements

I said it in the acknowledgements for my last book *Frazzled* and I'll say it again for this book, no one helped me write this besides my editors Joanna and Ed, the monk – Gelong Thubten, and the neuroscientist – Ashish Ranpura. Besides them, I thank my publisher Venetia Butterfield from Penguin for publishing and my agent, Caroline Michel from PFD, for agenting, and for both women being generally fabulous. Outside of them no one, not even a mouse, showed up to help me write this. I have only myself to thank.

Otherwise, outside of me, I'd like to thank the myriad of unknowing authors whose information I borrowed (as we say in the book biz). I hope they don't get too upset. If any of you authors are reading this, whatever I used of yours, please feel free to change back into your own words.

Anthony W. Bateman and Peter Fonagy: *Handbook of Mentalizing in Mental Health Practice*

Brene Brown: *I Thought It Was Just Me: Women Reclaiming Power and Courage in a Culture of Shame*

Dean Burnett: *The Idiot Brain: A Neuroscientist Explains What Your Head is Really Up To*

Mine Conkbayir: *Early Childhood and Neuroscience: Theory, Research and Implications for Practice*

Alistair Cooper and Sheila Redfern: *Reflective Parenting: A Guide to Understanding What's Going on in Your Child's Mind*

Louis Cozolino: *The Neuroscience of Human Relationships: Attachment and the Developing Social Brain*

Louis Cozolino: *The Social Neuroscience of Education: Optimizing Attachment and Learning in the Classroom*

Matthew Crawford: *The World beyond Your Head: How to Flourish in an Age of Distraction*

Susan David: *Emotional Agility: Get Unstuck, Embrace Change and Thrive in Work and Life*

Joe Dispenza: *Evolve Your Brain: The Science of Changing Your Mind*

Robin Dunbar: *How Many Friends Does One Person Need?: Dunbar's Number and Other Evolutionary Quirks*

Robin Dunbar: *The Human Story*

Mark Epstein: *Thoughts without a Thinker: Psychotherapy from a Buddhist Perspective*

Charles Fernyhough: *The Voices Within: The History and Science of How We Talk to Ourselves*

Piero Ferrucci: *The Power of Kindness: The Unexpected Benefits of Leading a Compassionate Life*

Luc Ferry: *A Brief History of Thought: A Philosophical Guide to Living*

Elaine Fox: *Rainy Brain, Sunny Brain: The New Science of Optimism and Pessimism*

Barbara L. Fredrickson: *Love 2.0: Finding Happiness and Health in Moments of Connection*

Clive Gamble, John Gowlett and Robin Dunbar: *Thinking Big: How the Evolution of Social Life Shaped the Human Mind*

Eugene T. Gendlin: *Focusing: How to Gain Direct Access to Your Body's Knowledge*

Sue Gerhardt: *Why Love Matters: How Affection Shapes a Baby's Brain*

Paul Gilbert: *The Compassionate Mind (Compassion Focused Therapy)*

Paul Gilbert and Choden: *Mindful Compassion*

Linda Graham: *Bouncing Back: Rewiring Your Brain for Maximum Resilience and Well-being*

Rohan Gunatillake: *This is Happening: Redesigning Mindfulness for Our Very Modern Lives*

Rick Hanson: *Buddha's Brain: The Practical Neuroscience of Happiness, Love and Wisdom*

Rick Hanson: *Hardwiring Happiness: How to Reshape Your Brain and Your Life*

Yuval Noah Harari: *Sapiens: A Brief History of Humankind*

Sam Harris: *Waking Up: Searching for Spirituality without Religion*

Bruce Hood: *The Self Illusion: Why There is No 'You' Inside Your Head*

Fred Hoyle: *The Intelligent Universe: A New View of Creation and Evolution*

Laura A. Jana: *The Toddler Brain: Nurture the Skills Today That Will Shape Your Child's Tomorrow*

Steven Johnson: *Mind Wide Open: Why You are What You Think*

Sue Johnson: *Hold Me Tight: Your Guide to the Most Successful Approach to Building Loving Relationships*

Jon Kabat-Zinn: *Full Catastrophe Living: How to Cope with Stress, Pain and Illness Using Mindfulness Meditation*

Jack Kornfield: *A Path with Heart*

Thomas Lewis, Fari Amini and Richard Lannon: *A General Theory of Love*

Beau Lotto: *Deviate: The Science of Seeing Differently*

Dr Shanida Nataraja: *The Blissful Brain: Neuroscience and Proof of the Power of Meditation*

Kristin Neff: *Self-compassion*

Wes Nisker: *You Are Not Your Fault and Other Revelations: The Collected Wit and Wisdom of Wes 'Scoop' Nisker*

Adam Phillips and Barbara Taylor: *On Kindness*

Michael Puett and Christine Gross-Loh: *The Path: A New Way to Think about Everything*

Marshall B. Rosenberg: *Living Non-violent Communication: Practical Tools to Connect and Communicate Skillfully in Every Situation*

Tamara Russell: *Mindfulness in Motion: Healthier Life through Body-centred Meditation*

Sharon Salzberg: *Lovingkindness: The Revolutionary Art of Happiness*

Daniel J. Siegel and Mary Hartzell: *Parenting from the Inside Out*

Daniel J. Siegel and Dr Tina Payne Bryson: *The Whole-brain Child: 12 Proven Strategies to Nurture Your Child's Developing Mind*

Tania Singer and Matthias Bolz (eds.): *Compassion: Bridging Practice and Science*

Robert Skidelsky and Edward Skidelsky: *How Much is Enough?: Money and the Good Life*

Shawn T. Smith: *The User's Guide to the Human Mind: Why Our Brains Make Us Unhappy, Anxious and Neurotic and What We Can Do about It*

L. S. Vygotsky: *Thought and Language*

Peter C. Whybrow: *American Mania: When More is Not Enough*

Mark Williams, John Teasdale, Zindel Segal and Jon Kabat-Zinn: *The Mindful Way through Depression: Freeing Yourself from Chronic Unhappiness*

D. W. Winnicott: *The Child, the Family and the Outside World*

Anthony Wolf: *I'd Listen to My Parents if They'd Just Shut Up: What to Say and Not Say When Parenting Teens*

Lewis Wolpert: *Malignant Sadness: The Anatomy of Depression*

And a special mention to Andrew Dellis, postdoctoral Fellow at the Research Unit in Behavioural Economics and Neuroeconomics at the University of Cape Town.

And to all the many many other experts I milked for information for this book who shall remain nameless, mainly because I forgot their names, but thank you anyway.

Index

Index

Index

Index

Index

Index